SETTING ENVIRONMENTAL STANDARDS

SETTING ENVIRONMENTAL STANDARDS

GUIDELINES FOR DECISION-MAKING

Edited by

H. W. de Koning

Division of Environmental Health
World Health Organization
Geneva, Switzerland

WORLD HEALTH ORGANIZATION
GENEVA
1987

ISBN 92 4 154214 4

TYPESET IN INDIA
PRINTED IN ENGLAND
86/7112–Macmillan/Clays–4500

Contents

Contents

Preface

Awareness about environmental pollution has increased during recent years so that there is now widespread appreciation of the serious health risks and the need for preventive measures. Such measures, implemented voluntarily or through legislation, have many technical and economic ramifications, as well as social and public health implications.

The various aspects of the decision-making process for the development of environmental standards are discussed in this publication from the standpoint of health. Several sections deal with the definition of appropriate health objectives or norms while other sections discuss the strategies and legislative instruments available to achieve these norms. In practice, of course, the decision-making process must be adapted to the overall economic, social, public health, and planning policies of each country. An attempt has been made throughout the text to provide examples and illustrations of how individual countries have done this.

Various procedures related to the decision-making process are described to provide the reader with an insight into and an understanding of what is involved and what uncertainties may surround the information that is being used. More detailed information on these procedures can be obtained from the references cited.

Many experts in different fields have contributed to the preparation of this publication. Some have prepared whole sections, while others have provided comments and suggestions for improvement, participated in meetings or acted as consultants. A list of contributors is given on pp. vi–vii and the contribution of all those involved is gratefully acknowledged. Special mention should be made of the work of Dr W. Muir, Hampshire Research Associates Inc., Alexandria, VA, USA, who, particularly in the early stages of the project, played a significant role in its coordination.

This book is the result of cooperation between the World Health Organization (WHO) and the Environmental Law Centre, which is located in Bonn, Federal Republic of Germany, and forms part of the Secretariat of the International Union for Conservation of Nature and Natural Resources (IUCN). .The Centre provided support in the preparation of this publication and technical advice throughout the project. The United Nations Environment Programme provided financial support for the publication of this book.

Comments and queries regarding this publication should be addressed to the Division of Environmental Health, World Health Organization, 1211 Geneva 27, Switzerland.

The views expressed in this book should not be construed as representing either decisions or policies of the International Union for Conservation of Nature and Natural Resources, the United Nations Environment Programme, or the World Health Organization.

Contributors and reviewers

V. Aalto, World Health Organization, Brazzaville, Congo

M. Baram, Franklin Pierce Law Center, Concord, MA, USA

C. Bartone, Pan American Center for Sanitary Engineering, Lima, Peru

M. Bothe, University of Hanover, Hanover, Federal Republic of Germany

W. Burhenne, IUCN Commission on Environmental Policy, Law and Administration, Bonn, Federal Republic of Germany

F. Burhenne-Guilmin, Environmental Law Centre, Bonn, Federal Republic of Germany

K. Buštueva, Central Institute for Advanced Medical Training, Moscow, USSR

J. Butler, Environmental Protection Agency, Washington, DC, USA

J. Carmichael, United Nations Industrial Development Organization, Vienna, Austria

E. Clark II, The Conservation Foundation, Washington, DC, USA

C. Cochrane, International Centre for Industry and Environment, Villeneuve de Berg, France

G. Davila, World Health Organization, Washington, DC, USA

B. Dieterich, World Health Organization, Geneva, Switzerland

E. Engstrom, National Food Administration, Uppsala, Sweden

S. Epstein, University of Illinois, Chicago, IL, USA

S. Fluss, World Health Organization, Geneva, Switzerland

M. Forster, Environmental Law Centre, Bonn, Federal Republic of Germany

M. Gilbert, United Nations Environment Programme, Geneva, Switzerland

B. Gillespie, Organization for Economic Cooperation and Development, Paris, France

D. Gushee, Congressional Research Service, The Library of Congress, Washington, DC, USA

R. Harcourt, Department of Health, Wellington, New Zealand

J. Huismans, United Nations Environment Programme, Geneva, Switzerland

O. Ikeda, Environment Agency, Tokyo, Japan

F. Irwin, The Conservation Foundation, Washington, DC, USA

D. Jayasuriya, World Health Organization, Geneva, Switzerland

F. Kaloyanova, Institute of Hygiene and Occupational Health, Sofia, Bulgaria

S. Kato, Ministry of Health and Welfare, Tokyo, Japan

M. Key, University of Texas, Houston, TX, USA

V. Kodat, Ministry of Health, Prague, Czechoslovakia

O. Kolbasov, Institute of State and Law, Moscow, USSR

W. Kreisel, World Health Organization, Manila, Philippines

R. Luken, Environmental Protection Agency, Washington, DC, USA

Z. Madar, Institute of State and Law, Prague, Czechoslovakia

U. Marinov, Environmental Protection Service, Jerusalem, Israel

H. Mattos de Lemos, United Nations Environment Programme, Nairobi, Kenya

C. Millar, International Atomic Energy Agency, Vienna, Austria

W. Muir, Hampshire Research Associates Inc., Alexandria, VA, USA

M. Nakamura, WHO Western Pacific Regional Centre for Promotion of Environmental Planning and Applied Studies, Kuala Lumpur, Malaysia

S. Niu, Institute of Health, Beijing, China

G. Ozolins, World Health Organization, Geneva, Switzerland

D. Popescu, University of Bucharest, Bucharest, Romania

P. Recht, Brussels, Belgium

E. Rehbinder, J. W. Goethe University, Frankfurt am Main, Federal Republic of Germany

W. Rowe, American University, Washington, DC, USA

P. Royabhorn, National Environment Board, Bangkok, Thailand

H. Salas, Pan American Center for Sanitary Engineering, Lima, Peru

M. Saric, Institute for Medical Research and Occupational Health, Zagreb, Yugoslavia

M. Sheikh, World Health Organization, Alexandria, Egypt

E. Smith, World Health Organization, Geneva, Switzerland

R. Smith, Department of Health, Canberra, Australia

K. Snidvongs, National Environment Board, Bangkok, Thailand

M. Sommers, Department of National Health and Welfare, Ottawa, Canada

M. Stephenson, Economic Commission for Europe, Geneva, Switzerland

P. Stief-Tauch, European Economic Commission, Brussels, Belgium

K. Symon, Charles University, Prague, Czechoslovakia

S. Tabacova, Institute of Hygiene and Occupational Health, Sofia, Bulgaria

R. Tardiff, Environ Corporation, Washington, DC, USA

G. Vettorazzi, World Health Organization, Geneva, Switzerland

P. Waight, World Health Organization, Geneva, Switzerland

R. Waller, Department of Health and Social Security, London, England

R. Zielhuis, University of Amsterdam, Amsterdam, The Netherlands

Chapter 1

Introduction

For this publication, environmental pollution is defined as energy or waste materials that are discharged into the environment where they can cause damage to human health (Holdgate, 1979). This definition excludes potentially hazardous materials used by individuals on themselves, for example, cosmetics, food additives, pharmaceuticals, or tobacco. The toxicity of environmental pollutants depends on their formulation and concentration. For example, the organic combinations of mercury, especially methylmercury, pose a greater hazard than the inorganic mercury salts. Similarly, high concentrations of sulfur dioxide (SO_2) in urban areas can undoubtedly harm both plants and people, even though sulfur is an essential component of all living organisms.

Pollutants may reach man through different pathways, e.g., via water, air, food, or consumer products, but one source usually contributes the major proportion of the pollutant. This pathway is called the "critical pathway". If the environmental health effects of a substance are to be monitored and controlled in the most efficient and economic way, this critical pathway must be identified. If this is not possible, the quality and quantity of total exposure via multiple pathways must be carefully assessed. The absence of such knowledge is one of the principal obstacles to the control or prevention of the adverse health effects associated with environmental pollutants. For example, a child suffering from lead poisoning may have been ingesting lead in drinking-water that flows through lead pipes, in paint or soil containing lead, or in food contaminated by lead through the air, water, soil, or via the food chain. When setting ambient air standards or permissible industrial effluent standards for lead, therefore, it is necessary to consider whether the amounts of lead reaching the child from all other sources should be taken into account.

An additional problem is that some pollutants are chemically very stable and degrade only very slowly. This stability means that the pollutant persists in the environment resulting in its gradual accumulation, for example, in the soil. A well known example of this type of pollutant is the insecticide DDT, but there are many other examples.

For some years there has been a considerable development of environmental policies at the national, regional, and international levels (Schaefer, 1981). The scope of these policies includes both the reduction of pollution and the preservation of natural resources, as well as the promotion of an improved quality of life. The control of environmental pollution, and in particular the setting of standards, raises difficult problems for governments, because different people, or groups of people, have different views of the extent to which a government should protect its citizens from risk and at what cost. Often it is a question of distribution: who is bearing the risk[1] and who the cost? (Technical Information Project, 1979).

For example, the people living near a waste dump are more likely to suffer the consequences of an accident at the site than people from another area who use its services as a treatment or disposal facility. On the other hand, the immediate neighbours of a chemical manufacturing plant may not oppose its installation, even if the risk involved in the factory siting is similar to that of the waste dump, because they will benefit economically from its presence through the creation of jobs.

It is particularly important to encourage broad participation in the decision-making process since the risk involved is often not evenly distributed; different groups may be particularly affected by environmental pollution for various reasons such as age or nutritional status, as well as for geographical reasons. In addition, the way in which people balance costs and benefits will differ. A major challenge of environmental standard-setting for governments and citizens is learning to communicate with each other about risks and to make decisions on the basis of information that is often incomplete.

Toxicity is the capacity of a substance to cause injury to a living organism. Hazard refers to the potential of a pollutant to induce harm (World Health Organization, 1977). The purpose of environmental health standards is to reduce or eliminate health or environmental hazards.

The assessment of the pollution hazard should be a strictly scientific process—a matter of evaluating probabilities using the best available information about the dispersion of a pollutant and the associated effect on the health of man or other targets. Once the likelihood of a particular level of effect under particular circumstances of emission and dispersal has been determined, value judgements become important; the effects on the socioeconomic system must be considered. The socioeconomic system in turn determines the types of effect that are acceptable, and the measures that may be taken to control the pollution. In practical terms, this control will involve limitation of effluents or emissions of pollutants, of radiation to the environment, or of the exposure of individuals. Another important consideration is prevention of hazard; this is a fundamental factor in the organization of a system of protection for man.

It is becoming increasingly evident that both prevention and control of environmental pollution involve a number of complex societal

---

[1] *Risk* is the probability of unfavourable or undesirable effects appearing as a result of a given exposure.

systems and must be analysed within a framework that is sufficiently broad to include all intrinsic and extrinsic factors that might affect human health and other targets (Davos & Nienberg, 1980). Each society will have a particular legal and procedural framework for making decisions about control measures and for determining whether these should be advisory or mandatory.

The main steps in the formulation of public policy decisions to protect health and welfare from environmental hazards usually occur in two stages, as follows.

1. Scientific stage

(a) *Knowledge of the hazard*—involves identification and character-ization.

(b) *Evaluation of the risk*—establishes the probability and severity of potential adverse effects on health and safety.

(c) *Assessment of hazard* —determines routes of exposure and es-timates the number of people exposed.

At the conclusion of this stage it should be possible to determine the levels of pollutants that do not produce adverse effects and to establish necessary safety margins, thus establishing goals or norms for national environmental pollution prevention and control programmes.

2. Political and administrative stage

(a) *Determination of acceptable risk*—views problem not as a scientific matter, but rather one of opinion.

(b) *Determination of public to be protected*—considers not only healthy individuals but also population groups whose particular physio-logical make-up or state of health need to be taken into account.

(c) *Consideration of human ecology*—sees man in balance with his environment.

(d) *Choice of control technology*—requires both formulation of strategy and selection of appropriate control techniques.

(e) *Legislation/standards* —considers existing national legal framework and identifies necessary legal strategies.

(f) *Economics* —strikes a balance between costs and benefits.

This stage requires knowledge of the technical, social, financial, legal, and institutional implications of the solutions to be adopted. This knowledge promotes the examination of links between environmental problems, the solutions, and society (Rodricks & Tardiff, 1984; Males, 1985). At this stage, consideration is given to the means of achieving the environmental health goals.

The framework outlined above is somewhat schematic and in most cases it will only be followed in a generalized way. It puts forward a logical progression in decision-making that, in "real life" situations, might not always be followed. Also, it proposes a decision-making process that is "science-driven". In practice, the decision-maker might have to find solutions for existing environmental problems, in which case the process might work in reverse: from the political to the scientific stage (see also Chapter 2). Lastly, although the process outlined above is linear, in practice it is cyclical, as constant improvements in information are likely to result in adjustments and changes being made to strategies, standards, and methods. However, no matter how the process operates, the elements described remain valid; they are discussed further in the other chapters of this book.

This approach to decision-making is not without its critics. Criticisms stem in part from the inadequacy of the data on which the assessment is based. For example, uncertainties concerning the effects of chronic exposure to environmental pollution make it almost impossible to quantify the associated risks. Further criticism stems from the concept of risk acceptability—acceptable to whom?

Faced with such uncertainties some would suggest that it might be better to take all practicable steps to eliminate avoidable risk, irrespective of precise quantification. However, the prevailing approach to pollution control is based on the principle that with scarce resources, attempts should be made to relate the control imposed to the hazards of exposure, and this approach is followed in this publication.

Chapter 2

Identification of priority pollution issues

There are generally insufficient resources available to deal with all the pollution problems in a country, and it will be necessary to establish priorities. Two general criteria should be considered in doing so (Whyte & Burton, 1980):

1. The boundaries of the problem must be defined. For example, a decision has to be made as to whether the risk to human health is the sole or major criterion for control. In the past, priority ratings have tended to focus on human health alone but increasingly, hazards to animals, plants, and natural areas have prompted environmental action in their own right.

Within human health and well-being, a hierarchy of effects can be identified from minor temporary ailments through acute illness to chronic diseases. Particular problems arise with chronic diseases, which are often difficult to relate to specific hazards or sources of risk. For others the significance of effects is uncertain. For example, it is known that at levels of less than 100 µg of lead per 100 ml of whole blood, anaemia does not usually occur. However, these low concentrations of lead in the blood affect the activity of an enzyme, porphobilinogen synthase (EC 4.2.1.24) (see Fig. 1 and 2, pages 18 and 19). Can this be considered to be a significant effect on human health? The question of what constitutes a health effect is discussed further in Chapter 4.

2. The problem in question must be put into a wider context by consideration of other risks and/or benefits. Risks can be evaluated in terms of the additional hazard they present over:

—what occurs naturally in the environment;
—what has been tolerated for long periods of time with no apparent ill effects;
—the level that is accepted as beneficial (e.g., in the case of pesticides).

For example, natural background levels have been used as a yardstick in measuring the risk associated with nuclear power production, the potential risk of adding fluoride to the domestic water supply as a public health measure, and the assessment of elevated noise levels near airports and traffic routes.

The question arises as to how national agencies should choose which pollutants must be controlled from among the thousands introduced into the environment. Traditionally, this decision has been based on a subjective consideration of a number of factors such as immediate hazard, public concern, feasibility of control, etc.; formal procedures have not generally been used. However, most developed countries and international organizations have now adopted systems for setting priorities in order to provide more rigorous guidance. Five criteria (World Health Organization, 1976) are usually applied in determining the extent to which a pollutant may pose an environmental hazard. These are:

—Severity and frequency of observed or suspected adverse effects on human health. Of importance are irreversible or chronic effects, such as genetic, neurotoxic, carcinogenic, and embryotoxic effects including teratogenicity. Continuous or repeated exposure generally merits a higher priority than isolated or accidental exposure.
—Ubiquity and abundance of the pollutant in the environment. Of special concern are inadvertently produced chemicals and substances that add to a natural hazard.
—Persistence in the environment. Pollutants that resist environmental degradation and accumulate in man, in the environment, or in food chains, deserve attention.
—Environmental transformation or metabolic alterations. Since alterations may produce chemical substances that have greater toxic potential, it may be more important to ascertain the distribution of the derivatives than that of the original pollutant.
—Population exposed. Attention should be paid to exposure involving a large proportion of the general population, or occupational groups, and to selective exposures of highly vulnerable groups such as pregnant women, newborn children, the infirm or the elderly.

Selection of pollutants for control

In practice, hazard identification begins with one component of the problem, usually the source of the effect, and does not consider the system as a whole at the outset. Some methods of identification are systematic while others appear to be more or less *ad hoc*. This is a pragmatic response to the different ways in which hazards are discovered. In the following paragraphs the different reasons for which pollutants may be selected for attention by environmental decision-makers are discussed.

Systematic evaluations

In the United States of America, efforts have been made to identify priority chemical pollutants systematically by using a scoring system to rank each substance (Environmental Protection Agency, 1977). Several other efforts in different parts of the world rely upon the knowledge and opinions of groups of experts.

At the international level, the Council for Mutual Economic Assistance (CMEA) and the Organization for Economic Cooperation and Development (OECD) both have programmes on methods of setting priorities for testing of new and existing chemicals. As a result of cooperation among CMEA countries during recent years, a number of documents have been published including a reference book on problems of industrial toxicology (GKNT, 1986). This publication contains guidelines and recommendations for studying various aspects of the biological effects of chemical compounds as well as procedures for establishing sanitary standards in CMEA countries. OECD recently published a guidance document, in two volumes, for the selection for further testing of chemicals on which data are inadequate. The second volume of the report includes nearly 700 citations that are useful in setting priorities (Organization for Economic Cooperation and Development, 1984).

Local or foreign public health crises

Often, national authorities are forced into the decision-making process by the discovery of a major local problem. For example, the Japanese government in the 1960s had to react quickly to deal with the mercury poisonings at Minamata (Katsuma, 1968). After the experience at Minamata, other countries around the world attempted to assess their own local situation with regard to mercury, and to take appropriate action.

Environmental problems are rarely confined to one country and in most instances sooner or later a particular problem will occur in several locations. As a result, the setting of standards and priorities for environmental control in one country may be influenced by the action of others. For example, when scientists in the Netherlands in the early 1970s discovered that certain harmful organic substances were present in drinking-water as a result of chlorination, many countries took appropriate control action.

Research

In some cases, environmental health issues appear on the decision-making agenda as suspected problems. For example, several countries have taken action since 1975 either to limit their production capacity of chlorofluorocarbons (CFC) or to ban (or severely restrict) their use as aerosol propellants. These actions were taken on the basis of a theory that the release of chlorofluorocarbons into the atmosphere will eventually significantly reduce the amount of ozone in the stratosphere that shields the earth's surface from harmful solar ultraviolet radiation (Council on Environmental Quality, 1975). While supporting evidence for this theory has been presented, the restrictions have been implemented before there has been any direct observation of a decrease in the ozone layer.

Outside opinion

Often, the press or local political figures will focus attention upon an issue to such an extent that priorities for action must be changed. In Canada, for example, press and politicians have drawn attention to the existence of several hundred miles of railroad bed made from asbestos tailings. As a result, environmental health officials have been obliged to consider the associated risks and possible remedial action.

Chemical similarity

Problems may often be suspected on the basis of the chemical similarity between one substance and another that is known to be hazardous. For example, polybrominated biphenyls quickly received regulatory scrutiny around the world after the chemically similar polychlorinated biphenyls were shown to constitute a major environmental hazard.

The factors discussed above may influence local and national authorities in their efforts to set priorities for action. Alternatively, such authorities may wish to devote their full effort to controlling hazards and may therefore rely upon various international programmes to establish priority lists. These lists may be supplemented by local and national surveys to pinpoint specific local problems.

Advantages of international cooperation

The advantages of international cooperation include the availability to decision-makers of expert reviews of information from many countries at low cost. Generally, such reviews provide not only internationally relevant suggestions for priority action but also associated information on evaluation of health effects, suggested safe levels for human, plant, and animal exposure, production or emission data, and concentrations in different environmental media, etc. Further information on relevant documentation produced by different international programmes is given in the following paragraphs.

WHO Environmental Health Criteria programme

Established in 1973, the main objective of this programme is to assess existing information on the relationship between exposure to environmental pollutants, or other physical factors, and human health, as well as to provide guidelines for setting exposure limits that are consistent with health protection. This programme was later incorporated into the newly established International Programme on Chemical Safety (IPCS).[1]

In view of the large number of environmental agents and factors that may adversely influence human health, the preparation of criteria documents must be based on clearly defined priorities. Each criteria document comprises an extensive scientific review of a specific environmental pollutant, group of pollutants, or physical factor(s); the information provided ranges from sources and exposure levels to a detailed account of the available evidence concerning effects on human health. Over fifty documents have been published to date. A list of Environmental Health Criteria publications is given in Annex 1.

International Register of Potentially Toxic Chemicals

The International Register of Potentially Toxic Chemicals (IRPTC) is part of the United Nations Environment Programme. The Register serves as an international data bank and information service on possible chemical hazards.[2] Its activities involve the development of data profiles on chemicals, the operation of a query-response service, and the regular publication of the IRPTC bulletin which contains up-to-date information on chemicals. The main objective of this programme is to facilitate access by countries to existing data on the effects of chemicals on man and the environment, and thus to contribute to a more efficient use of national and international resources. The programme also helps to identify the potential hazards of chemicals and pollutants, and to improve awareness of such hazards.

At the present time, the IRPTC has data profiles on more than 500 chemical substances; a list of these is available from IRPTC. Two specific files from the Register have been published separately: the *IRPTC legal file* (also accessible on-line) and *Treatment and disposal methods for waste chemicals*. Examples of pertinent records are given in Annex 2.

International Agency for Research on Cancer

In 1965, the World Health Assembly established the International Agency for Research on Cancer (IARC) in Lyon, France. One major activity of IARC over the last decade has been the publication of monographs evaluating the possible carcinogenic hazards from chemical substances and complex mixtures. Up to September 1986, 38 volumes had been issued, concerning approximately 700 chemicals, groups of chemicals, and industrial processes. Of these, 30 chemical substances, mixtures, or groups of products, and 9 industrial processes

---

[1] MERCIER, M. *The International Programme on Chemical Safety.* Unpublished WHO document, EHE/80.14, Rev. 1.
[2] *Summary of the Second Meeting of Experts on Listing of Environmentally Dangerous Chemical Substances and Processes of Global Significance, 21–25 November 1983.* Geneva United Nations Environment Programme, 1983.

have been found to be causally associated with cancer in man. In addition, 63 chemical substances or mixtures of chemicals and 5 industrial processes are considered to be probably carcinogenic to man. There are another 115 chemicals for which evidence of carcinogenicity has been found in animals, but for which no studies or data are available on their effects in man (Vainio et al., 1985). A listing of the IARC monographs is given in Annex 3.

FAO/WHO pesticide reviews

Joint meetings of the FAO Working Party of Experts on Pesticide Residues and the WHO Expert Committee on Pesticide Residues have been held regularly since 1965. Proceedings of the meetings are published in reports that summarize the conclusions reached, including values of recommended acceptable daily intake (ADI) and maximum residue limit (MRL), as well as the recommendations for further work on the evaluation of pesticide chemicals. The reports are supplemented by accompanying volumes containing summaries of toxicological and residue studies for each of the individual pesticides considered at the meeting (Vettorazzi & Radaelli-Benvenuti, 1982). A listing of compounds discussed, many of which have been reviewed several times during past years, is given in Annex 4.

FAO/WHO food additive reviews

Since 1957, toxicological evaluations of food additives and contaminants have been carried out by the Joint FAO/WHO Expert Committee on Food Additives (JECFA); the 29th report of this Committee was issued in 1986 (WHO, 1986b). These reviews provide guidance to the authorities that are responsible for setting standards and for organizing the control of foods. For each chemical substance evaluated, the relevant report recommends an acceptable daily intake (ADI). A list of additives and contaminants reviewed to date is given in Annex 5.

Guidelines for drinking-water quality

Since 1971, WHO has coordinated international reviews of scientific and epidemiological information on different pollutants to provide guidelines for drinking-water quality. The latest revision of these guidelines was published in two volumes in 1984 and 1985 (World Health Organization, 1984, 1985). These two publications provide detailed and comprehensive data for each pollutant, upon which the guideline values are based. A table summarizing the WHO water quality guidelines is given in Annex 6.

Air quality guidelines

The WHO Regional Office for Europe has developed guidelines for air quality (Annex 7). The primary aim of these guidelines is to provide a basis for protection of public health from adverse effects of air pollution and for elimination or reduction of pollutants that are known or likely to be hazardous.

The guidelines are intended to provide background information and guidance to governments for making risk-management decisions—in particular, for planning control strategies and setting national or local standards. The guidelines should, however, be considered in the context of prevailing exposure and environmental, social, economic, and cultural conditions. Under certain circumstances, there may be valid reasons to pursue strategies that will result in pollutant concentrations above or below the guideline value. In this regard, a recent publication of the WHO Regional Office (World Health Organization, 1987) provides further information that would be of assistance in making the appropriate decision.

Workplace standards

Literally thousands of chemical substances are used in the workplace. Many countries have adopted occupational exposure limits for numerous airborne contaminants found in the workplace. The International Labour Office has prepared a compilation of these regulations (International Labour Office, 1980). Within the WHO programme on recommended health-based limits for occupational exposure to airborne contaminants (El Batawi & Goelzer, 1985), reports have been published on heavy metals (World Health Organization, 1980), selected organic solvents (World Health Organization, 1981), pesticides (World Health Organization, 1982), vegetable dusts (World Health Organization, 1983a), respiratory irritants (World Health Organization, 1984a), and mineral dusts (World Health Organization, 1986a). The limits set out in these reports are health-based, which means that only scientific and not economic evidence is considered concerning exposure levels and associated health effects. An overview of the methods used to calculate tentative safe exposure levels in the workplace in countries belonging to the Council for Mutual Economic Assistance (CMEA) was published recently (GKNT, 1986).

United Nations Environment Programme

The United Nations Environment Programme (UNEP) has prepared a list of environmentally dangerous pollutants and processes that are harmful at the global level, in an attempt to stimulate awareness among governments and the public of the hazardous effects of environmental pollutants on man and the environment (United Nations Environment Programme, 1982). This list will be revised periodically. Pollutants of global significance are defined as those that occur widely in the environment in significant quantities as a result of their transport through air, water, and food chains, or because they are present in commodities traded internationally on a large scale. The selection of

11

dangerous pollutants for the UNEP list is also based on the persistence and transformation of a substance in the environment, rates of bioaccumulation and biomagnification, the exposed populations, toxicity, exposure level, and effects on the physical and chemical environment.

The wide variety of external influences on the process of priority setting, and the pressure to deal with recently discovered problems as they arise, before others have been resolved, pose a dilemma for environmental decision-makers. On the one hand, they must be responsive to current needs, and, on the other hand, they must maintain sufficient consistency to allow the typically lengthy analytical and decision-making processes to be completed for as many pollutants as possible.

Chapter 3

Information on health effects

The toxic effects of pollutants found in the environment are commonly divided into broad classes, such as acute toxicity, chronic toxicity, carcinogenicity, teratogenicity, and mutagenicity. A substance may be capable of causing a range of toxic effects, including cancer.

The main distinguishing characteristic between these categories of effects is the assumption that dose-thresholds exist for the acute and chronic toxic effects, but that such thresholds do not exist (or have not been demonstrated) for carcinogenic effects. In the former case, standard-setting aims to achieve a total dose of the substance that is below the level at which any injury would result to any individual in the population. For pollutants assumed not to have a threshold level, it follows that some finite risk may exist at any dose level above zero; thus, standard-setting may aim to reduce exposure to zero (a level that is often neither measurable nor achievable in practice) or to specify a dose level that will contribute only a negligible increase in the lifetime risk to the individuals and/or the population exposed.

Studies of health effects

The first step in establishing environmental standards for a pollutant is to determine the nature of any hazards associated with it. Whether the national or local authority chooses to conduct appropriate studies itself, or decides to rely upon the expert opinion of others, it is important that the strengths and limitations of the data and information available are known.

Epidemiology[1]

Epidemiological studies on human populations can in principle provide definitive evidence of the health risks associated with a particular pollutant. However, for the study of environmental pollutants, their main weakness is that they are relatively ineffective in proving that observed health effects are the direct result of exposure to a particular substance. In practice, the

---

[1] Epidemiology is the study of the various factors that determine the occurrence and distribution of diseases and other physiological and pathological effects in human populations.

requirement for a pollutant to be present for a long time before discernible differences appear in a population militates against the use of epidemiology as a basis for regulatory decisions on environmental health issues. Unless the effect of the pollutant is unusual, highly specific, and recognizable by a physician—as in the case of angiosarcoma of the liver caused by exposure to vinyl chloride—it will pass unnoticed during a normal survey. For example, in order to detect a twofold increase in congenital malformations over a background rate of 2%, a total of 3000 births would have to be studied (Kline et al., 1977).

Different types of epidemiological study may differ considerably in their time-frames. In simple terms, cross-sectional or prevalence epidemiological studies refer to the present, retrospective or case-control studies to the past, while cohort and prospective intervention studies follow groups into the future. The limitations of such studies have recently been considered (World Health Organization, 1983). Some of the salient features of each type of epidemiological study are presented in Table 1.

Some examples where the setting of environmental health standards has been largely based on epidemiological studies are the fluoridation of drinking-water, air quality objectives for oxides of nitrogen, and the control of emissions of arsenic, asbestos, and vinyl chloride. In some instances, epidemiological studies can be used to confirm the upper limit of a risk estimate, e.g., in occupational exposure to ionizing radiation.

Clinical studies

Certain types of information about the effects of environmental pollutants can only be obtained by direct observations in man. For example, carefully controlled experiments on some types of effect, such as subtle changes in reaction time, behavioural functions, and sensory responses, can be carried out using low doses otherwise considered to be safe.

Animal studies

Acute animal studies are most commonly used to predict human response to short-term, high-level exposures, such as may occur following an accident; they can also provide a measure of the toxic potential of different compounds. Metabolic and pharmacokinetic studies are used to determine the absorption, distribution, and elimination of the test compound, its biotransformation and the rates at which these processes occur. The toxic effects of longer-term, low-level exposure are obtained from chronic or lifetime studies, as well as from subchronic studies of 28 days or longer. The rationale of such tests has been described in detail elsewhere (World Health Organization, 1982a).

Table 1. Major features of various study designs in environmental epidemiology

Study design	Population	Exposure	Health effect	Confounders are:	Problems	Advantages
Descriptive study	Various sub-populations	Records of past measurements	Mortality and morbidity statistics, case registries, etc.	Difficult to sort out	Hard to establish cause–result and exposure–effect relationships	Cheap, useful in formulating hypothesis
Cross-sectional study	Community or special groups; exposed vs. non-exposed groups	Current	Current	Usually easy to measure	Hard to establish cause–result relationship; current exposure may be irrelevant to current disease	Can be done quickly; can use large populations; can estimate extent of problem (prevalence)
Prospective study	Community or special groups; exposed vs. non-exposed groups	Defined at outset of study (may change during course of study)	To be determined during course of study	Usually easy to measure	Expensive and time consuming; exposure categories can change; high drop-out rate	Can estimate incidence and relative risk; can study many diseases; can infer cause–result relationship
Retrospective cohort study	Special groups such as occupational groups, patients, and insured persons	Occurred in past – need records of past measurements	Occurred in past – need records of past diagnosis and measurements	Often difficult to measure because of retrospective nature (e.g., past smoking habits)	Changes in exposure/effect over time of study; need to rely on records that may not be accurate	Less expensive and quicker than prospective cohort study, giving similar response, if sufficient past records are available
Time-series study	Large community with several million people; susceptible groups such as asthmatics	Current, e.g., daily changes in exposure	Current, e.g., daily variations in mortality	Often difficult to sort out	Many confounding factors, often difficult to measure	Useful for studies on acute effects
Case-control study	Usually small groups; diseased (cases) vs. non-diseased (controls)	Occurred in past and determined by records or interview	Known at start of study	Possible to eliminate by matching for them	Difficult to generalize due to small study group; some incorporated biases	Relatively cheap and quick; useful for studying rare diseases
Experimental (intervention) study	Community or special groups	Controlled/known	To be measured during course of study	Can be measured; can be controlled by randomization of subjects	Expensive; ethical consideration; study subjects' compliance required; dropouts	Well accepted results; strong evidence for causality

A methodology has been proposed for assessment of chronic toxicity on the basis of the dose–time relationship (Pinigan, 1976; Pinigan & Grigorevskaya, 1978). This method, widely used in the USSR for setting maximum permissible concentrations (MPC) for air pollutants, is based on the determination of dose–time relationships in one-month animal studies using relatively high concentrations. The established threshold concentration is subsequently extrapolated for longer period(s) of exposure, e.g., up to 4 months.

Most regulatory decisions on the toxicity of pollutants are based on the results of animal tests using mammalian species. Experiments with animals raise the fundamental question: how valid is the comparison between man and animals? Although the basic biological processes of molecular, cellular, and organ function are similar from one mammalian species to another, there are marked differences between the standard animal models and man.

Extrapolation from animals to man has been most successful for certain effects such as carcinogenicity, acute toxicity, and metabolic interference, as well as for deafness, behavioural changes, and influence on the immune system. The greater our understanding of the basic mechanisms of toxicity and interspecies differences, the more accurate will be predictions based on animal toxicology studies.

Short-term tests

The time, expense, and logistic difficulties involved in the conduct of animal tests have stimulated the search for short-term test methods to detect toxicity (World Health Organization, 1978). Further motivation for such research has also been provided by a growing concern for animal welfare. Mutational effects (induction of changes in genetic material) have been widely used as an index of toxicity. Such screening will not only detect substances that may induce birth defects, but will also give an indication of possible carcinogenicity, although the extent of the association between mutagenicity and carcinogenicity appears to depend greatly on the class of pollutant as well as on the properties of the test system used to assay mutagenic potential (Ashby, 1983).

There are over 50 mutagenicity tests in existence (National Research Council, 1983), of which the *Salmonella* assays first developed by Ames et al. (1973) are the best known, most widely used, and most thoroughly validated. In spite of their imperfections, these tests do provide a valuable tool for the detection of potential mammalian genotoxic carcinogens and mutagens. However, it should be borne in mind that, with the possible exception of the action of electrophilic agents, the multistage nature of carcinogenesis is too complex to be reduced to simple systems.

Structure–activity relationships

From the early days of pharmaceutical research, it has always been hoped that knowledge of the physicochemical characteristics of a substance could be used to predict its biological activity. Much information has been collected for various classes of compounds on the correlation between chemical structure, in terms of functional groups and spatial orientation, and parameters of toxicity (Birge, 1983). Short-term methods for predicting toxicity and maximum permissible concentrations for occupational and ambient air pollutants have been developed on the basis of such studies (Zaeva, 1964; World Health Organization, 1975; Nikiforov et al., 1979).

Dose–effect relationships

The dose received can either be expressed as the *total dose*, integrated over time, or as the *actual dose*, that is the amount present in the target organ at a point in time. The actual dose in the critical organ, which is often correlated with daily absorbed dose, is usually the most relevant measure with regard to graded effects, i.e., effects that can be measured on a graded scale of intensity; the *magnitude* of such an effect is directly related to the dose.

The total dose is more important with regard to quantal effects, i.e., those for which the *occurrence* of an effect depends on the absorbed dose. The term quantal means that the effects permit no gradation and can be expressed only as "occurring" or "non-occurring", i.e., there is no dose-threshold below which the effect will not appear, but the probability of experiencing the effect increases with increasing dose (World Health Organization, 1978). Hereditary effects and cancer caused by certain chemicals are considered to be quantal effects.

Observed effects are evidence of biological changes in an organism. A dose–effect[1] relationship can be seen in terms of either the increasing severity of one effect with increasing dose (e.g., eye irritation versus oxidant concentration in the atmosphere) or a series of effects, from the less severe to the more severe, with increasing dose. The latter case is illustrated in Fig. 1 and 2, which show the relationship between lead concentration in blood and various health effects in adults and children.

The terms effect and response are often considered to be interchangeable, but the latter refers more specifically to the proportion of a group of individuals that demonstrates a defined effect.

This section considers the different approaches to the development of standards for both carcinogens and substances thought to be noncarcinogens, with threshold (graded) or non-threshold (quantal) dose–effect relationships. One of the problems inherent in the establishment of a dose–effect relationship is that of extrapolating from the results of high doses observed during experimental studies to those of the low dose levels that are more typical of human exposure.

---

[1] The term exposure may be used in preference to dose; it is the amount of the agent taken up by the body per unit time over a period of time (World Health Organization, 1980).

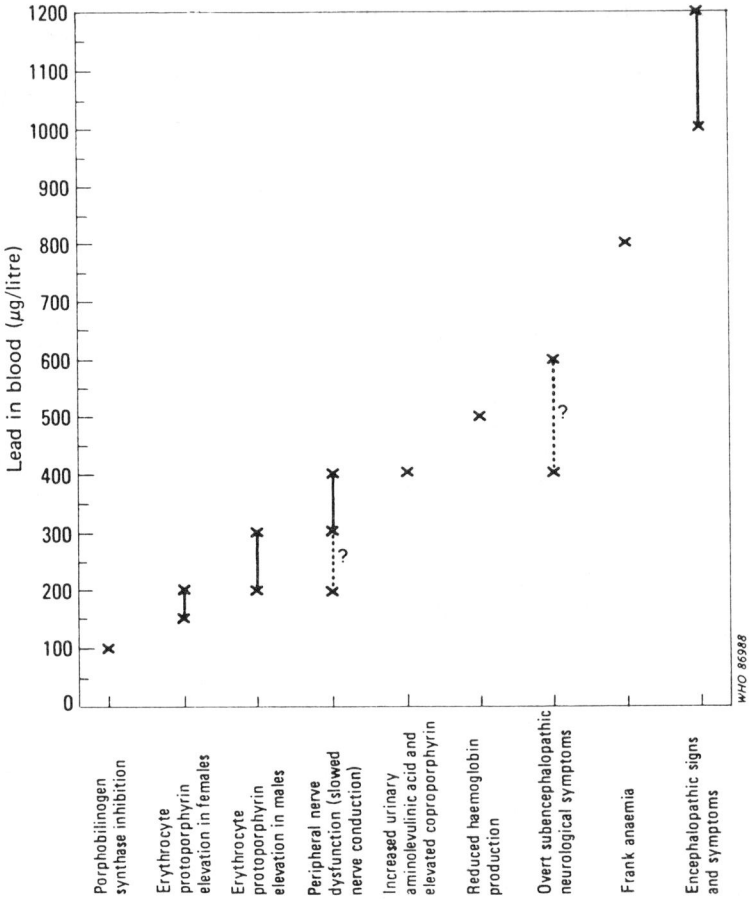

Fig. 1.   Lowest observed levels for lead-induced health effects in adults

Threshold effects

Different types of dose–effect curve are shown in Fig. 3. Curve 1 represents the simplest case, where there is no risk until a certain level of exposure is reached, at which point curve 1 leaves the abscissa

A more complex and more common situation is illustrated by curve 2. Here there are some effects at low doses, and these effects increase relatively slowly with increasing exposure, until a take-off point is reached after which the effects increase dramatically. This example illustrates the case where a few susceptible members of a population are affected by low-level exposure but the mass of the population remains unaffected until a certain threshold exposure or take-off point is reached.

Both curves 1 and 2 may be directly linked to policy options. In the case of curve 1, it would be appropriate to keep exposure of the

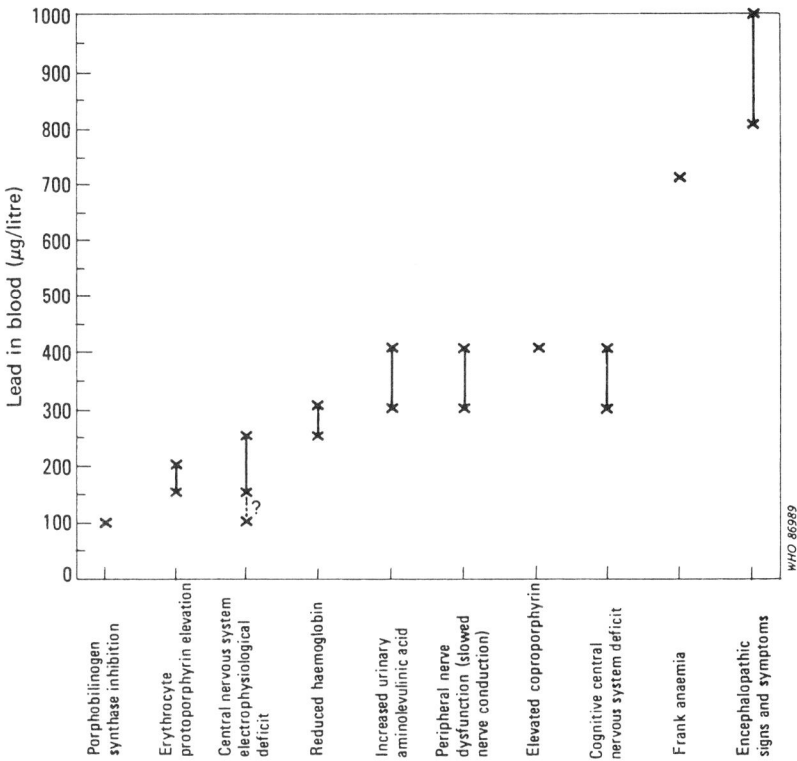

Fig. 2.   Lowest observed levels for lead-induced health effects in children

population below the threshold level. For curve 2, it may not be practicable to reduce exposure to zero in order to protect the relatively small number of susceptible people affected at low doses. In this case, the standards may be set at or somewhat below the take-off point, and additional steps taken to safeguard or reduce the exposure of the susceptible persons.

A more complicated situation is illustrated by curve 3 where the effects of exposure are impossible to separate from similar effects that occur from background exposure. In other words, there are some "effects" at zero dose and any dose above zero will increase those effects.

In some instances, a gradual increase in the intensity of the effect can be observed with increasing exposure (dose); an example is the effect of lead on enzyme systems connected with haem biosynthesis. One of the enzymes affected is porphobilinogen synthase (PS), the activity of which is known to decrease with increasing dose of lead. Using the blood lead levels as a measure of the dose, a negative linear relationship can be established between the dose and the logarithm of PS activity in the erythrocytes. Almost complete inhibition of the

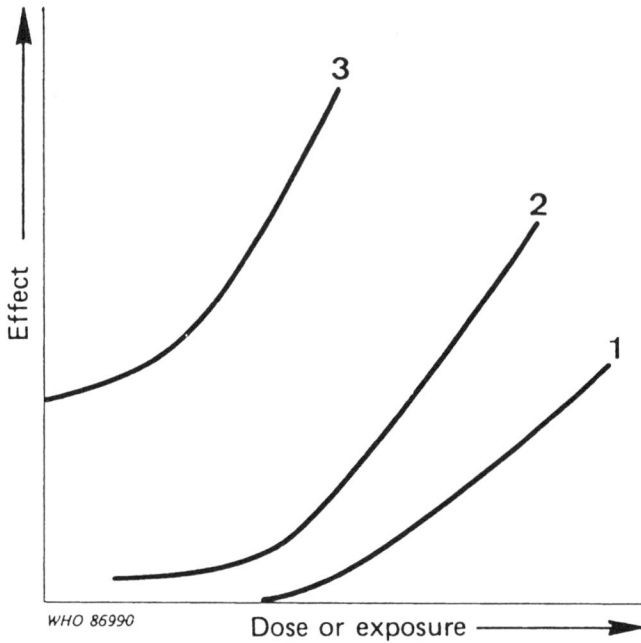

Fig. 3.   Illustration of different types of dose–effect curve showing a threshold

enzyme can be observed at blood lead levels that exceed 600 µg per litre. Demonstrable PS inhibition can often be observed at blood lead levels below 100 µg per litre.

In practice, it is not always possible to specify with confidence the slope of the dose–effect curve or to state exactly where any threshold level is. A number of factors account for this lack of precision. The vulnerability of individuals varies and physiological diversity in human populations is such that effects may vary according to the section of the population exposed. Measuring techniques have their limits of detection and monitoring of levels can be carried out in only a few selected samples or sites. At very low exposures, the effects may not be easily detectable; equally, data relating to the upper end of the curve are difficult to obtain because massive exposures are relatively rare.

The principal use of dose–effect curves is, therefore, to predict the consequences of very high and very low exposures.

Fig. 4 is a generalized exposure–effect curve showing how effects at the lower levels of exposure may be estimated by extrapolation from "middle range" observations.

The solid line to point A is the dose–effect curve, determined by a multiple-dosing experiment. Point A is the "no observed effect" level (NOEL) in mg per kg of body weight per day for the most sensitive adverse end point, as determined from a chronic multiple-dose animal

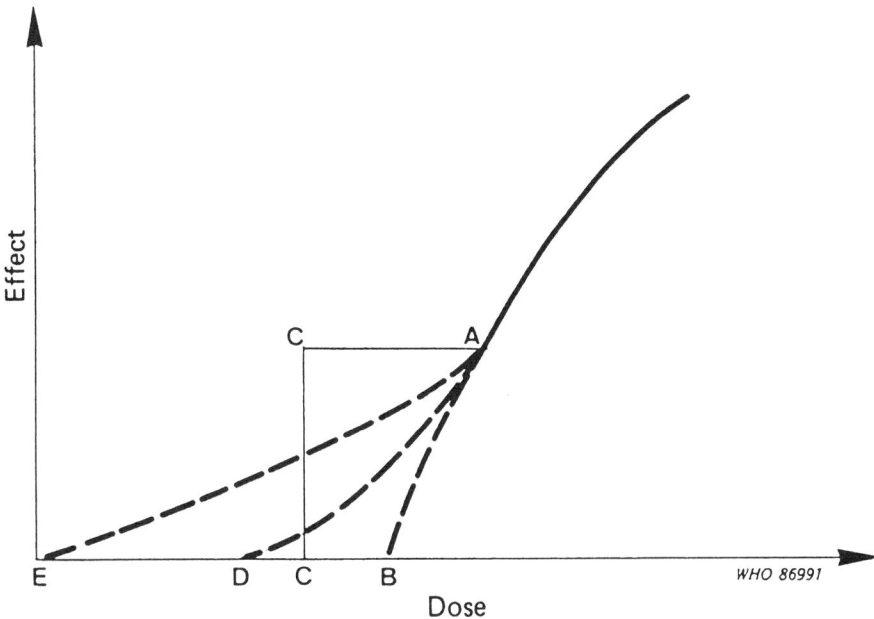

Fig. 4. Exposure–effect curve showing the different possible estimates at lower dose levels

study. Curves AB, AD, and AE are possible dose–response curves at lower doses, with points B, D, and E the respective thresholds for the adverse effect in the human population. In setting an acceptable daily intake (ADI) concentration (point C), a selected safety (uncertainty) factor is applied to the dose at point A. If the curve AB is the true dose–response curve, then the calculated ADI value will be lower than the threshold dose, thus indicating that the safety factor was appropriate. However, if AD or AE is the true dose–effect curve then the calculated ADI will be too high, and the safety factor too small; in this case, some individuals in the population will suffer adverse effects. The size of the gap between points C and B is also of interest, because if it is large, expenditure on control methods could be greatly in excess of what is needed.

Safety factors and extrapolations

In qualitative terms, the toxic action of environmental pollutants in man can usually be inferred, with a reasonable degree of certainty, from animal studies. On the other hand, the accuracy and reliability of the quantitative prediction of toxicity in man depend on a number of conditions, such as the animal species, design of the experiment, and methods of extrapolation used (World Health Organization, 1982a).

21

There are two fundamental problems of extrapolating animal data to man: (1) the problem of extrapolating projected response from one species to another, and (2) the problem of extrapolating from a tested dose–response range to a different dose range of practical importance.

The safety (or uncertainty) factor reflects the degree of uncertainty that must be incorporated in the extrapolation from experimental data to the human population. When the quality and quantity of dose–response data are high, the safety factor is low; when the data are inadequate or equivocal, the safety factor must be higher. However, it should be noted that when permitted doses of carcinogens are the practical equivalent of zero, safety factors are not especially relevant. In the case of carcinogens, the term safety factor should be replaced by the term "acceptable risk" (see pp. 24–27).

The following general guidelines have been adopted by the National Academy of Sciences (NAS) Safe Drinking-Water Committee and are also used by the Environmental Protection Agency in the USA in the development of drinking-water standards (National Academy of Sciences, 1977):

> 10 factor—Applied to data from valid experimental studies on prolonged human intake, with no indication of carcinogenicity. This 10-fold factor protects the sensitive members of the population.
>
> 100 factor—Applied when experimental results from studies of human intake are not available, or are scanty; valid results of long-term intake studies on one or more species of experimental animals; no indication of carcinogenicity.
>
> 1000 factor—Applied when there are no long-term or acute human data; scanty results on experimental animals; no indication of carcinogenicity.

These are not rigid rules and should be applied using scientific judgement in each particular situation.

In the USSR the term "reserve coefficient" is used to represent the difference between the maximum permissible concentration (MPC) and the threshold level for physiological responses. The "reserve coefficient" may be viewed as the safety factor that is incorporated in each standard. Several approaches have been proposed in the USSR for the derivation of safety factors. According to Izmerov (1973), the maximum permissible concentration should include a safety factor of 30 % in all cases. This means that if a presumed threshold is found, the next level tested is 30 % lower. If there is no pollutant-induced activity at that level, the experimentation can stop, with that final concentration tested becoming the maximum permissible concentration. This may seem a very small safety factor when compared to the traditional factors of 10 or 100 adopted in the United States, but it must be remembered that the USSR uses this method to prevent the occurrence of the most sensitive indicator of exposure and not necessarily the pathological response (see also pp. 28–29).

Example of a threshold effect: para-dichlorobenzene

An example of a calculation of acceptable daily intake showing the use of exposure–effect data and safety factors is described below for *para*-dichlorobenzene.[1] Animal studies with various doses of *para*-dichlorobenzene have shown effects including liver and kidney damage, porphyria, pulmonary oedema, and changes in spleen weight. Human exposure to high concentrations of the chemical has been reported to result in pulmonary damage and haemolytic anaemia. A one-year gavage study in the rabbit using groups of 5 animals dosed with between 0 and 1000 mg of *para*-dichlorobenzene per kg of body weight per day resulted in weight loss, tremors, and liver pathology. The highest no-observed-adverse-effect level (NOAEL) was 357 mg of *para*-dichlorobenzene per kg per day. A subchronic study indicated a no-observed-adverse-effect level of 150 mg of *para*-dichlorobenzene per kg of body weight in the rat exposed by gavage. Animals received doses of 37.5, 75, 150, 300 or 600 mg of *para*-dichlorobenzene per kg per day in corn oil, 5 days per week for 13 weeks. No significant differences were observed in food consumption or gain in body weight compared with controls for either sex at any dose. At the two highest doses, an increase in the incidence and severity of renal cortical degeneration was detected microscopically.

Using this last experiment as the basis for calculations with an additional factor reflecting that the exposure in the experiment occurred for only 5 days each week, the provisional ADI was computed as follows:

$$\text{ADI} = \frac{150\,\text{mg/kg of body weight/day} \times 70\ \text{kg/person} \times \frac{5}{7}}{100 \times 10}$$

$$= 7.5\,\text{mg per person per day}$$

where 100 is the safety factor appropriate for use with a no-observed-adverse-effect level from animal studies with no comparable human data; and 10 is an additional safety factor because the duration of exposure in the experiment was significantly less than a lifetime.

Minimal data were available on the contributions of food and air to exposure, so an arbitrary designation of 20% was chosen as the maximum allocation from drinking-water. If the daily water intake per person is assumed to be 2 litres per day, then the allocated ADI for water (AADI) is:

$$\text{AADI} = \frac{\text{ADI} \times \text{water allocation}}{2\ \text{litres/day}}$$

$$= \frac{7.5\,\text{mg/day} \times 20\%}{2\ \text{l/day}} = 0.75\,\text{mg/l}$$

[1] COTRUVO, J. A. Risk assessment and control decisions for protecting drinking-water quality. In: *Evaluation of methods for assessing human health hazards from drinking-water.* Lyon, France, International Agency for Research on Cancer, 1986 (Internal Technical Report No. 86/001).

Non-threshold effects

The curves shown in Fig. 5 all represent possible non-threshold relationships. Curve 1 is the classical linear exposure–effect relationship when no threshold exists. Zero risk occurs only at zero exposure. Curve 2 is a variant showing reduced sensitivity to risk at lower levels, and curve 3 is the reverse showing increased sensitivity at lower levels of exposure (Whyte & Burton, 1980).

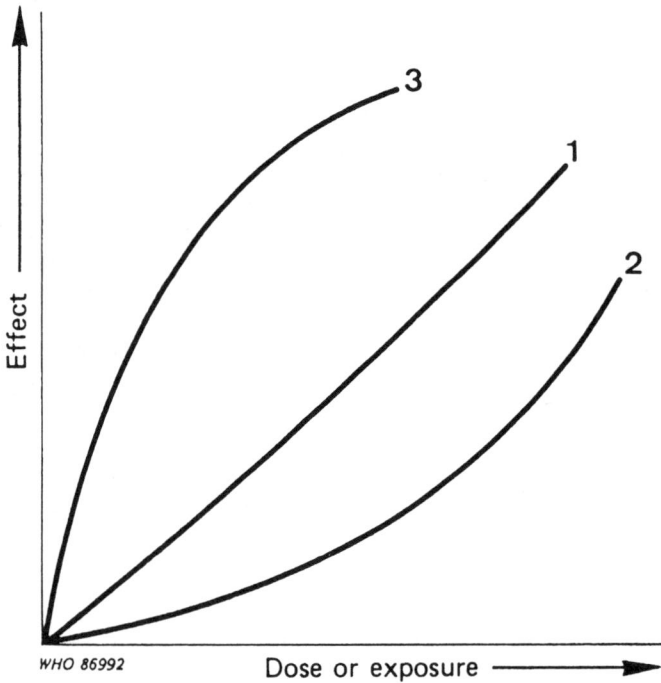

Fig. 5. Illustration of different types of dose–effect curve showing zero threshold

The existence of a threshold for cancer is a subject of contention among risk assessors. The argument against a threshold assumes that a single point mutation of the DNA can lead to the uncontrolled growth of a somatic cell that eventually produces cancer. A probabilistic argument can be made that, while an individual may have a threshold, different individuals may have different thresholds or may have none at all. The arguments for a threshold are based on the existence of gene repair mechanisms, immune defences, and epigenetic mechanisms.

The US National Academy of Sciences Safe Drinking-Water Committee has outlined four principles that it believed should be useful in dealing with the assessment of hazards involving chronic irreversible toxicity or the effects of long-term exposure (National Academy of Sciences, 1977). These were intended to apply primarily to cancer risk from pollutants that cause somatic mutation. These principles, discussed below, may also be applicable to mutagenesis and teratogenesis.

1. Effects in animals, properly qualified, are applicable to man. This premise underlies all of experimental biology, but it is often questioned with regard to human cancer. Virtually every form of human cancer has an experimental counterpart, and every form of multicellular organism is susceptible to cancer. There are differences in susceptibility among animal species, different strains of the same species, and individuals of the same strain. However, extensive data indicate that substances that are carcinogenic to animals are likely to be carcinogenic to man, and vice versa.

2. Methods do not exist, at present, to establish a threshold for the long-term effects of toxic agents. It is not possible to establish experimentally carcinogenesis thresholds that would be applicable to a total population. There is no scientific basis for the estimation of "safe" doses of carcinogens using classical exposure assessment techniques. Even experimental bioassays involving a large number of animals are likely to detect only powerful carcinogens. Negative results in such bioassays do not prove that the agent is unequivocally safe for man.

3. The exposure of experimental animals to high doses of toxic agents is a necessary and valid method of assessing possible carcinogenic effects in man. There is no choice but to use a small number of animals relative to the size of the exposed human population, and then to use biological models to extrapolate the results in order to estimate risk at low doses. An incidence as low as 0.01 % would represent a risk to 20 000 people in a population of 200 million, whereas the lower limit of reproducibility in common animal studies would be an incidence of 10 %. The best method available, at present, is to assume a direct proportionality between dose and tumour incidence, with no threshold. It is possible that the actual risk to man is greater than predicted by small animal studies because of the longer human lifetime and hence exposure period.

4. Material should be assessed in terms of human risk rather than as "safe" or "unsafe". Decisions must take account of possible benefits as well as risks, and should consider the nature, extent, and recipient of these benefits. It is often necessary to accept risks when there are substantial benefits, but risks imposed on persons who gain no benefit are generally not acceptable. Also, in view of the nature of cancer, the long latent period of its development, and the irreversibility of chemical carcinogenesis, it would be highly improper to expose the general population to an increased risk if the benefits were small, questionable or restricted to particular segments of the population. Such benefit–risk considerations not only must be based on hard scientific evidence but also must be ethical, with as broad a population base as possible being used in the decision-making process.

Quantitative assessment of cancer risk relies on dose–response functions to establish the likelihood that cancer will occur in individuals exposed to low doses of pollutants (Ricci & Molton, 1985). The process of cancer formation provides biologically plausible bases for some of the functions. Quantal functions describe whether the adverse effect is present or not. Time-to-response functions account for the differences in longevity among exposed individuals. Such models as the one-hit, the probit, the logistic, the multihit, and the multistage do not explicitly model those dependencies.

The one-hit model assumes that a single hit causes irreversible damage to the DNA (point mutation) and leads to cancer. In the multistage model, a cell-line must pass through several (k) stages before a tumour is irreversibly initiated. The rate at which cell-lines pass through these stages is a function of the dose rate. In the multihit model, k dose-related hits to the sensitive tissue are required to initiate a cancer. The Weibull model includes the more reasonable assumption that these hits must occur in a single cell-line and that different cell-lines compete independently in producing a tumour.

In all of these models it is assumed that the rate at which dose-related hits occur is a linear function of the dose rate. The most important difference between the multistage and the Weibull and multihit models is that in the Weibull and multihit models all hits must result from the dose, whereas in the multistage model, passage through some of the stages can occur spontaneously. The practical implication of this is that the Weibull and multihit models predict a lower risk at low doses than does the multistage model.

The remaining models—probit and logistic—are not derived from mechanistic assumptions about the cancer process. They may be thought of as representing the distribution of tolerance to the carcinogenic insult in a large population. There is no evidence to support these particular distributions over others and, in some respects, all the models discussed may be interpreted as tolerance distributions.

Similar principles for assessment of carcinogenic hazards are used in member countries of the Council for Mutual Economic Assistance. Setting of maximum permissible concentrations for carcinogens in the ambient air is based on the experimental determination of the probability of occurrence of carcinogenic effect for a period exceeding the normal lifespan in laboratory animals and subsequent extrapolation to man using a reserve coefficient (Yanyševa, 1972).

Example of a non-threshold effect: determination of water quality guidelines for carcinogenic substances

In 1984, the World Health Organization established guidelines for biological quality, aesthetic quality, and levels of radioactivity, inorganic chemicals, and organic chemicals in drinking-water (World Health Organization, 1984). Of the 20 organic substances for which guideline values were set, 10 were considered as potential carcinogens (Table 2).

Table 2.   Water quality guidelines for substances considered to be potential carcinogens

Constituent	Guideline value (μg/l)
benzene	10
benzo[a]pyrene	0.01
carbon tetrachloride	3
chloroform	30
1,2-dichloroethane	10
1,1-dichloroethene	0.3
hexachlorobenzene	0.01
tetrachloroethene	10
trichloroethene	30
2,4,6-trichlorophenol	10

Guidelines for carcinogens were set where reliable data were available from two animal species, preferably supported by muta- genicity data or population studies. Tentative guideline values were recommended for three pollutants—carbon tetrachloride, tetrachloro- ethene, and trichloroethene—where it was felt that the carcinogenicity data available did not justify definitive guidance.

It was assumed that thresholds for carcinogenicity were either non- existent or non-measurable. A multistage model was used in many cases with a data base similar to the one used for the US Water Quality Criteria (Environmental Protection Agency, 1976) and calculated values were generally rounded off to the nearest digit. The guidelines state: "The model is designed to estimate the highest possible upper limit of incremental (excess over background) risk from a lifetime of exposure to a particular daily amount of substance." An "acceptable" risk of 1 in 100 000 per lifetime was arbitrarily selected as the criterion.

Since the guideline values for the carcinogens were computed from a conservative, hypothetical, mathematical model that could not be experimentally verified, they should be interpreted with caution. The uncertainties involved are considerable and a variation of about two orders of magnitude (i.e., from 0.1 to 10 times the value) could exist.

Chapter 4

Assessment of exposure

Assessment of exposure to a given environmental pollutant involves, in the first instance, a consideration of the probable sources of the substance. Such sources may be industrial, domestic, consumer products or any number of others. These sources are not mutually exclusive; for example, asbestos fibres may be introduced into the environment from mining and the production of asbestos products, and from the use of asbestos in brake linings, water pipes, etc. A good inventory of sources will provide much information on critical pathways, possible environmental concentrations, and the hazard posed to the general population or subgroups at particular risk (World Health Organization, 1982b).

Assessment of exposure also involves making environmental measurements to determine the average and peak concentrations of the pollutant in the relevant media—air, water, and food. The aim of these measurements is to establish the levels in the different media and to study how these vary with time and place. This information should permit an estimate to be made of the exposure of the general population and subpopulations at particular risk.

Data obtained from monitoring together with information on lifestyle factors can be used in models to estimate exposure. These models may be very simple, indicating, for example, that 10% of the population is exposed to concentrations over a certain value. More complex models might identify people living in a certain geographical area as being exposed above a certain preset limit. The ultimate objective here is to establish the relative number of people who will react to the pollutant by manifesting a certain effect.

High-risk groups

Once the toxic dose for the "normal healthy" population has been determined, consideration must be given to high-risk groups such as infants and young children, the elderly, pregnant women and their fetuses, the nutritionally deprived, the physically debilitated, individuals with genetic disorders, and those exposed to excessive amounts of the pollutant. These groups must be quantified to determine the proportion of the population at greater risk from a certain pollutant than the so-called "normal" population.

Because of their physical vulnerability and immaturity, fetuses, infants (especially premature infants), and young children are at greater risk than adults. For example, several Japanese children born to mothers exposed to methylmercury in fish in Minamata suffered congenital malformations even though the mothers showed few or no symptoms of mercury poisoning (World Health Organization, 1976).

Nutritional deficiency may also increase susceptibility to the adverse health effects of some pollutants. Inadequate protein in the diet of rats increases the toxic effects of most pesticides, but decreases, or fails to alter, the toxicity of other agents. Vitamin E deficiency increases ozone toxicity, whereas vitamin E supplementation decreases ozone toxicity in rats. Dietary deficiencies of calcium and iron, quite common in the United States of America, significantly potentiate the toxicity of lead.

Individuals who are suffering from some disease may also show increased susceptibility to pollutants. Impairment of the functional capacity of the excretory systems, e.g., a renal disorder, can change the rate of clearance of pollutants and their toxic metabolites from the organism, and thus modify its effects. Impaired liver function can change rates of metabolic conversion, particularly detoxification of certain pollutants or their excretion in bile. Also, those suffering from cardiovascular or respiratory disease are at greater risk from the effects of carbon monoxide (CO) or sulfur dioxide (SO_2) than are healthy individuals.

It is very important to know which individuals are at high risk with respect to pollutants, because they will be the first to experience morbidity and mortality as the level of the pollutant increases. A standard that protects the high-risk sections of the population will also protect the rest of the population. Consequently, information concerning both the number and distribution of high-risk groups should play an integral role in the derivation of standards for pollutants in both ambient and industrial air as well as in drinking-water and food.

A theoretical comparison between normal and high-risk segments of the population with regard to the onset of toxic effects is shown in Fig. 6. As pollutant levels increase, adverse health effects will be seen first in the high-risk group. The precise difference in sensitivity between a statistically "normal" individual and one in the high-risk subpopulation varies with the specific cause of the high-risk condition. In any case, it is probably not entirely correct to assume that a well-defined threshold for an adverse effect exists within the highly diverse human population, although such an assumption may be of practical significance in many cost–benefit analyses. Neither can it be said that there are distinct thresholds for the normal and the high-risk groups.

If sufficient data are available on the levels at which particular effects occur in high-risk groups, an acceptable standard may be the level that does not result in subclinical physiological changes of a potentially damaging nature. Alternatively, it may be necessary to apply a safety factor to the dose that is toxic to the general population, in an effort to protect susceptible groups.

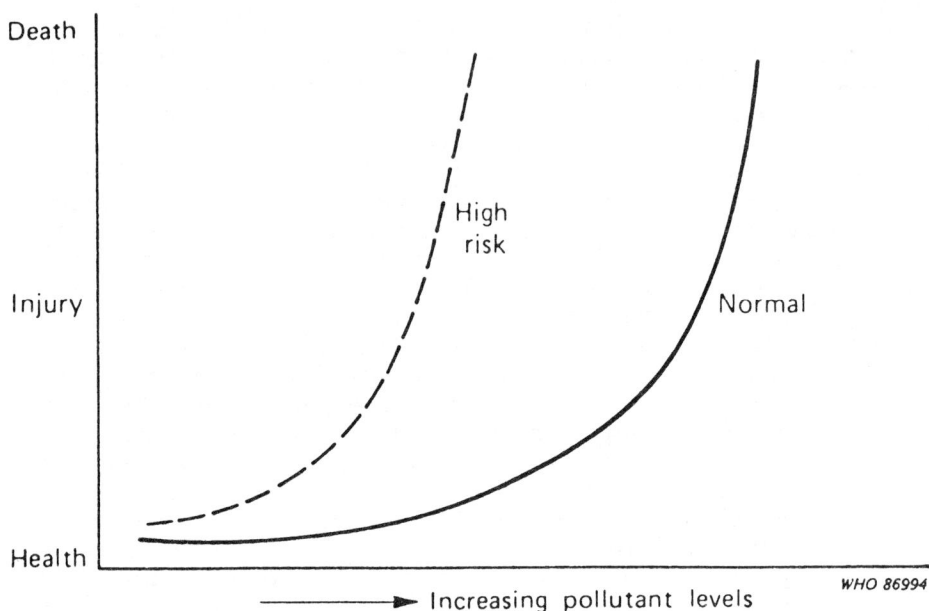

Fig. 6. Comparison of the response of high-risk and normal individuals to increasing pollutant levels

Definition of adverse health effects

In setting a standard for the control of an environmental pollutant, the effect that the population is to be protected against must be defined (World Health Organization, 1977). A hierarchy of effects on health can be identified, ranging from acute illness and death through chronic and lingering diseases, minor and temporary ailments, to temporary emotional or behavioural changes (Table 3). Should the first evidence of a biochemical or physiological change in an individual be considered

Table 3. Range of effects on human health of exposure to environmental pollutants

Premature death of many individuals
Premature death of an individual
Severe acute illness or major disability
Chronic debilitating disease
Minor disability
Temporary minor illness
Discomfort
Behavioural changes
Temporary emotional effects
Minor physiological change

adverse? Or is the appearance of clinical symptoms necessary for the response to be defined as adverse?

One school of thought considers a biological response, of whatever level of intensity, as evidence either of an impending detrimental effect on health, or as a direct expression of injury (World Health Organization, 1975). Consistent with this is the idea that any measurable concentration of the pollutant in body tissues, however low, is evidence of excessive exposure, since the body is stressed in its attempt to metabolize, store, or excrete it. In contrast, others assume that the homoeostatic and compensatory mechanisms in the body make it possible to offset such insults so that many toxic pollutants can be dealt with by the body without any threat to health.

Fundamentally, the approach to standard-setting depends largely on the accepted concept of the term "state of health". In the case of most environmental and industrial standards in the USA, for example, it is assumed that there is no threat to health as long as the exposure does not induce a disturbance of a kind and degree that overloads the normal protective mechanisms of the body. In marked contrast, in the USSR, a potential for ill health is said to exist as soon as the organism undergoes a detectable change, of whatever kind, from its normal state. In the USSR, maximum permissible concentrations for environmental pollutants are set below the level that causes physiological and other changes of uncertain significance, i.e., below the level that produces adaptive and compensatory changes in the most vulnerable groups (Buštueva, 1976). Thus, the concepts of tolerance limits or thresholds of response are markedly different in these two countries, at least with respect to what indicates an effect (see Table 4). Other countries generally follow one of these two approaches.

Table 4. Contrasts in approaches to standard-setting in the USA and the USSR[a]

USA	USSR
1. Minor physiological adaptive changes are permitted	1. Maximum allowable concentration will not permit the development of any disease or deviation from normal
2. Economic and technological feasibility are important considerations in the development of standards	2. The principle is that standards should be based entirely on health and not on technological and economic feasibility
3. Values are time-weighted averages	3. Concentrations are maximum values
4. Research emphasis is on pathology	4. Research emphasis is on nervous system testing
5. Except for carcinogens, goals of near zero exposure are not widely adopted	5. The goal is a level of exposure that does not strain the adaptive and compensatory mechanisms of the body (Ryazanov, 1961; Buštueva, 1976)

[a] Adapted from Calabrese, 1978.

In spite of these differences in approach, the trend in standard-setting in many countries, including the USA, has been to lower the permissible levels as a result of the development of increasingly more sensitive indicators of preclinical, physiological, biochemical, and other indices of functional disturbance. One example of this relates to lead; there is now agreement and concern among scientists that relatively low blood lead levels (in the range of 400–600 µg of lead per litre) in children may increase the risk of learning impairment very soon thereafter.

Chapter 5

Strategies for prevention and control

Once it has been decided that exposure to an environmental pollutant needs to be prevented or reduced in order to protect human health, the next step is to select appropriate measures to achieve this goal. Exposure can be avoided if the pollutant is prevented from being released or if it is degraded naturally or treated before it reaches its target. Reduction or control measures may be taken at various points on the path between the source of the contaminant and its target—a person or resource (see Fig. 7).

Frequently, measures will be taken against a pollutant at more than one point. For some contaminants, such as heavy metals, control actions may be introduced in nearly every area. For example, the European Economic Community has at least 17 directives that apply to cadmium, including limit values for the discharge of cadmium into water from six manufacturing processes (von Moltke et al., 1985).

The decisions made concerning the points of intervention and the type and level of control to be used must take into account many factors, including the chemical and physical characteristics of the pollutant, its transport through the environment, and possible related exposure. Such decisions are also influenced by the technology available and the financial resources of both the government and industry. The legal and cultural traditions of the country also play a role. Control at each stage has its own advantages and disadvantages. The opportunities and constraints faced, particularly by developing countries, are discussed in Chapter 8. The discussion below outlines different possible control strategies, using cadmium as an example where possible.[1]

Source-oriented measures

Product design, marketing, and use

For pollutants that are produced or used in the manufacture of a product, the first point on the pathway between the source and the target individual or population is the design and marketing of the

---

[1] The examples involving cadmium, unless otherwise referenced, are taken from von Moltke et al., 1985, Annex 3 (*Measures to control cadmium in selected countries*).

34

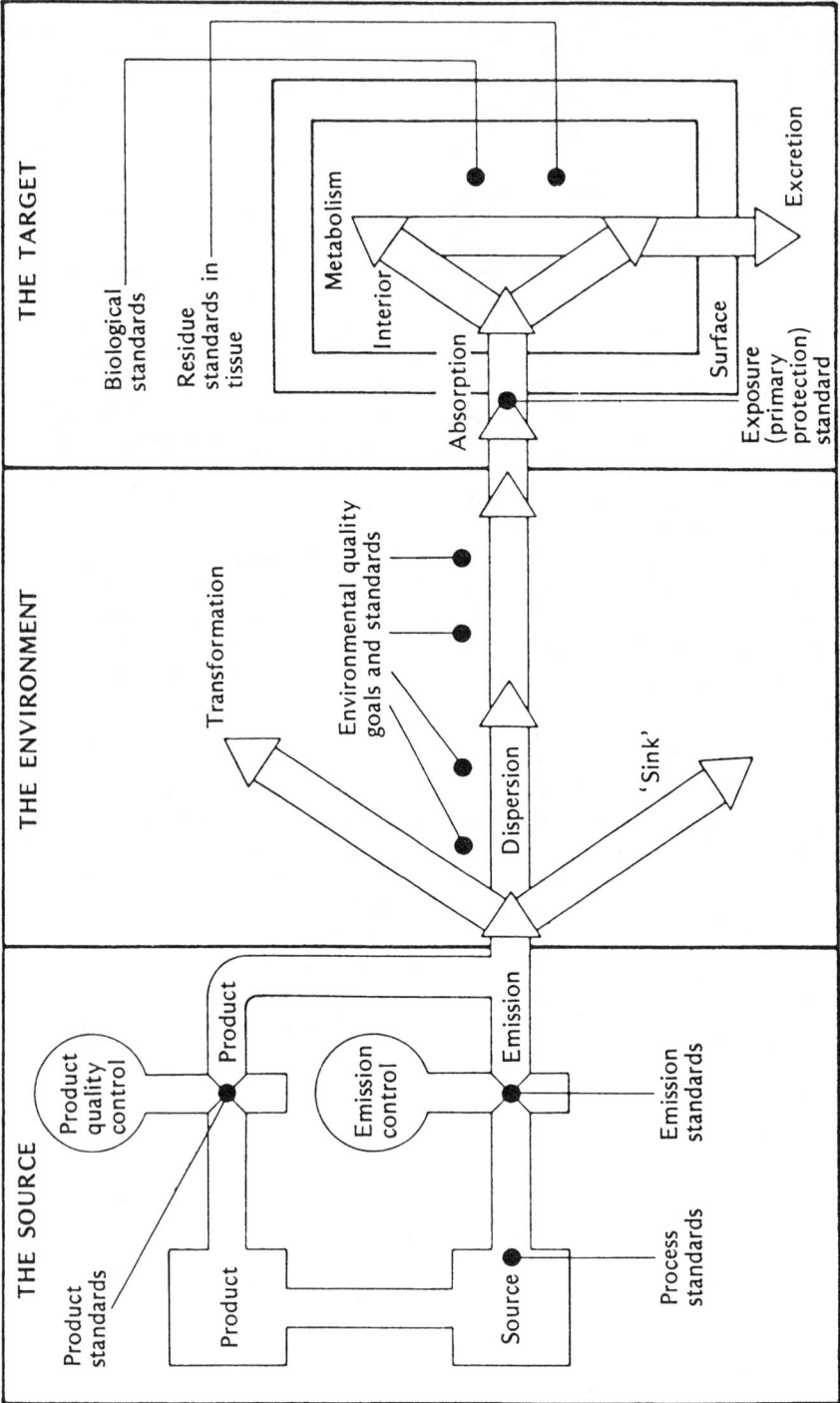

Fig. 7. The pollutant pathway showing possible points at which standards may be set (from Holdgate, 1979).

product itself. If the substance does not have a known threshold level and there is thus no safe level of exposure, it makes sense to start at this stage to see if it is possible to prevent completely its release, e.g., by redesigning the product, or searching for a substitute.

Governments can influence product design in a number of ways. One way is to ban the product or the use of a substance for certain purposes. Sweden has banned the use of cadmium except in certain areas—electroplating, pigments, and stabilizers for plastics. It may also be used in soldering if the product is appropriately marked, but not on products that could come into contact with drinking-water or food. Another possibility is for the government to encourage the use of substitutes. In Finland, for example, the use of organic compounds instead of cadmium-based pigments is recommended. Research is being carried out in the Federal Republic of Germany on possible alternatives to the use of cadmium in plastics and paints, as well as on the possibility of recycling certain products, i.e., returning them to the manufacturer for safe disposal. These types of provision do not involve the setting of any levels for cadmium. The approach of setting levels has been taken by some countries for paint; the United Kingdom, for example, sets a maximum concentration of 100 mg/kg (100 ppm) in dry film for paints on toys and the same concentration of soluble cadmium for crayons, brushes, and similar articles.

Another approach that can be used at this stage is the imposition of labelling and other handling requirements. In the European Economic Community, paints, inks, varnishes, adhesives, and similar products containing cadmium concentrations higher than 0.1 % by weight (1 g per kg) must comply with special requirements on type and strength of packaging material and on labelling identifying the product distributor. This is also the approach being used for cadmium batteries in a number of countries. In this case, in combination with recycling, such measures reduce exposure to a minimum since release of cadmium is unlikely during routine use.

Process of manufacture

Like the product design stage, the process stage offers an opportunity for reduction of the amount of a pollutant that enters the environment through the use of different methods of production. Control at this stage is likely to require an investment in research and development, but it may also reduce production costs. Control may affect individual companies differently because of variations in their needs and circumstances. Some companies fear that their competitive advantage will be reduced if they reveal details of process changes. However, changes that reduce the quantity of waste products generated are being increasingly publicized (Huisingh & Bailey, 1982). Much work on production changes has been carried out under the heading of residuals management at the research institution, Resources for the Future, in the USA (Kneese & Bower, 1979).

Manufacturing process changes have been proposed in the Federal Republic of Germany to reduce the amount of cadmium used and released during electroplating. These include changing the amount of cadmium in the electrolytes and improving the rinsing technology, by recirculating the rinse-water and also by treating it separately from other wastes. Recirculation of wastewater with concomitant removal of toxic substances is also mandatory in Czechoslovakia. The substitution of other materials for cadmium in electroplating, such as zinc and aluminium, is also being tried in some countries.

To encourage changes in manufacturing processes, governments can provide tax incentives and support information exchange programmes. They can indirectly promote changes by enacting and enforcing regulations concerning emission standards and waste management programmes. Process changes may also be mandated directly. Japan required the replacement of mercury cells by diaphragm cells for the production of chlorine and caustic alkali after first requiring a process change, i.e., the recycling of all water containing mercury.

Operating practices

The modification of operating practices focuses primarily on the handling of materials in ways that prevent accidental or continuing release and that deal with them when they occur. These procedures include maintenance measures to avoid spills and to clean them up, careful storage and segregation of different types of waste so that they can be more easily recycled or treated, and thorough cleaning of equipment. Training is essential to ensure that operating practices are strictly adhered to. Recent accidents in the chemical industry in Bhopal, India, and West Virginia, USA, have highlighted the importance of such basic procedures in avoiding or mitigating serious exposure to chemicals and pollutants.

Methods for taking emergency containment and clean-up action and warning the surrounding community must be well developed to handle any accidents that do occur. Until recently, the development of these practices has been entirely the responsibility of the company, occasionally with assistance from trade associations. Under a European Economic Community Directive on industrial accidents (European Economic Community, 1982), companies using certain hazardous substances are now required to inform the appropriate government authority about their accident prevention and control plans. These plans must describe how substances are stored, used, and produced and the various risks involved. The responsible government agency reviews the company's plans and the adequacy of emergency off-site planning. The Environmental Protection Agency has developed guidance on emergency planning for communities (Environmental Protection Agency, 1985). This guidance includes a list of about 400 chemicals that are capable of producing serious health effects immediately upon or shortly after exposure.

Many countries have regulations concerned with the packaging, storage, handling, and disposal of toxic substances. In Romania, for example, pesticides may be used only by trained personnel, and may not be handled by persons under 18 years of age or by pregnant women.

Limits on releases into the air and water

Much of the effort in the environmental control of pollutants has been concentrated on the development of emission limits on industrial discharges to air and effluent limits on those to water. Recently, increased attention has also been paid to releases into the soil.

Emission or effluent standards may be expressed in terms of the permissible concentration of a pollutant in units of air emitted or wastewater discharged by a source, or in terms of the total load of pollutant per time unit, unit of production, or unit of energy or materials input. The prescription of number of samples to be tested, measurement methods, and statistical assessment procedures forms an integral part of the standard.

Emission or effluent standards may also be expressed in a more general way, e.g., in terms of significant danger to health or to the environment, or in terms of the possibility or feasibility of control measures. In this case, the emission or effluent standard is related to the preferred method of control. The great advantage of such a standard is its direct relevance to the polluting activity. Furthermore, it is a comparatively easy standard to implement, as far as the regulatory aspect is concerned.

A European Economic Community directive prescribing limits on cadmium discharges to water illustrates how such standards may be set (European Economic Community, 1983). Limits have been set for six industrial sectors at two levels, the first of which must be complied with by 1986, and the second by 1989. The limits are given in terms of milligrams of cadmium per litre of discharge water and grams of cadmium discharged per kilogram of cadmium handled. The standards cover: zinc mining; lead and zinc refining; the production of cadmium metal; manufacture of cadmium compounds; manufacture of pigments; manufacture of stabilizers; manufacture of batteries; and electroplating. Member states are required to set emission standards that are no less stringent than the most nearly comparable limit value, taking into account the best technical means available for control. Depending on the experience obtained with the directive, the Commission of the European Communities may propose more restrictive limit values in the future.

The directive also covers the manufacture of phosphoric acid and/or phosphatic fertilizer from phosphatic rock, but does not set limit values because no economically feasible technology exists for extracting cadmium from this rock. Member states are still expected to set emission standards for this process, however. A proposed strategy for the control of cadmium recommends that the Commission fund research on the development of processes to remove cadmium from phosphatic rock in the manufacture of fertilizer, and that a permissible level of cadmium content in fertilizer be specified. Thus emission levels need to be closely tied to process and product controls.

The European Economic Community directive controlling cadmium discharges into water also links emission standards to environmental quality objectives. Permitted concentrations in water bodies are those judged to be the maxima compatible with the protection of aquatic life. The concentration in sediments and shellfish is not to increase significantly over time. Methods to be used in monitoring cadmium levels are also specified.

In the USA, cadmium and its compounds are among 129 priority pollutants to be controlled in water. This is accomplished through limitations on effluents or setting of pretreatment standards according to industrial category. The industries covered include electroplating and metal finishing, production of non-ferrous metals, ore mining and dressing, porcelain enamelling, and manufacture of electrical and electronic components. The limits are enforced through a permit system.

Air emission standards have been set primarily in relation to energy production and incineration since these processes may release respirable particles containing cadmium. The Netherlands has a dust emission limit for coal-fired power plants, while the United Kingdom uses the "best practicable means" approach (see page 40). Numerous countries are looking at controls on various types of incinerator; the Netherlands requires wet scrubbers or semi-dry filtering with electrostatic precipitation.

Discharge standards for air and water have been a first step towards pollution control in many countries, often for particular industries, but in some cases, applied generally. Sometimes standards are more stringent for new plants than existing ones. The standards may be legally enforceable or they may be issued as guidelines, as is the case in India and Sweden. The United Kingdom operates on a case-by-case approach; although there is a list of presumptive limits for air pollutants, each case is judged individually. In the USSR, norms are set for maximum allowable emissions from every fixed source and every mode of transport. For fixed sources, a special permit is issued by the authorized state body, specifying the maximum allowable emission and detailing other conditions and requirements. Financial incentives can also play a role. In Hungary, air pollution emitters must pay fines if their level of emissions is above that considered acceptable.[1]

Emission levels, whether as guides or enforceable standards, are important in controlling pollution from industrial and combustion

---

[1] *Compendium of environmental guidelines and standards for industrial discharges.* Unpublished WHO document, EFP/83.49.

sources. They give both government and industry clear goals to aim for. Their development can, however, be extremely time-consuming and expensive for many industries. In addition, while they may encourage companies to change their products or processes, such standards have often led to the use of add-on pollution control equipment, such as filters, scrubbers or settling ponds. This equipment collects the pollutant rather than releasing it into the air or water in the immediate vicinity. The trapped pollutants are then often moved to another location and dumped on land. From this new location the pollutants that do not rapidly degrade move into the air or water at a later date. In the USA, cadmium was the tenth most frequently reported substance at 546 waste sites on the list for clean-up; it was detected at 82 sites (Environmental Protection Agency, 1984). It was found in the air at 31 sites and in the surface water and groundwater at 28 sites each. A number of cadmium waste streams that are now designated hazardous and thus require special handling may be the source of some of this cadmium. They include dust recovered from electric furnace production of steel and secondary lead-smelting control devices and wastewater treatment sludges from electroplating.

Because sludge from wastewater treatment plants is often applied to soil, there is now a need for the development of emission limits to the soil. The cadmium in this sludge usually comes both from industrial discharges to the treatment plant and non-point sources, and may cause a problem by accumulating at high levels in some foods. Sweden has set maximum concentrations in sludge used for agricultural purposes at 15 mg of cadmium per kg (dry weight). The Netherlands has imposed a limit for the amount of sewage sludge that can be applied to croplands of 2 tonnes per hectare per year and for grasslands of 1 tonne per hectare per year. The cadmium content must be below 10 mg/kg.

Treatment and disposal

Once a pollutant has been produced, it must either be degraded into forms that will not cause adverse health effects or be contained and isolated from the population. Some pollutants, including metals such as cadmium, do not degrade; once released into the environment, they continue to circulate in the air, water, and soil. Cadmium accumulates in the soil, from where it may move into the food chain, leach into the groundwater, erode into surface water, or be blown on the wind as dust. In order to prevent its build-up in the environment it needs to be contained as far as is possible by such measures as recycling of rinse-waters used during manufacturing processes involving cadmium. In Finland, 50–70% of wastewater from mines is recycled and the remaining cadmium is chemically precipitated. Closed water cycles are also used in basic metal and sulfuric acid industries. Other non-degradable pollutants need to be handled in a similar way.

Some contaminants can be treated by a wide range of biological and chemical methods that destroy their toxicity. Polychlorinated biphenyls, for example, can be destroyed chemically using sodium-based compounds that remove their chlorine atoms, or by high-temperature incineration. These processes are more expensive than landfill techniques, but they ensure that the compounds do not enter the environment at a later date. In the waters of the Great Lakes region of the USA and Canada, deposition from the air is apparently the largest source of polychlorinated biphenyls and some metals (Eisenreich et al., 1980). Other sources include rivers that are contaminated from landfills and then flow into the lakes. Thus the most important sources are no longer direct, but indirect discharges resulting from the lack of degradation or failure of containment processes.

Many countries have regulations that require careful handling of hazardous wastes. In the USA, land disposal of some wastes may be prohibited, giving further impetus to the use of treatment methods that destroy or contain the pollutant. A more direct approach is to require particular methods of treatment for specific materials. In the USA there are special requirements for the disposal of polychlorinated biphenyls and, under the Toxic Substances Control Act of 1976, the authorities can specify the type of disposal and containment to be used for a compound. There are also recommended procedures for the disposal of pesticides containing cadmium, and their containers (USA, 1984).

Best practicable means

This approach implies the use of the best available and economically feasible control technology. It is inherent in the term "feasible" that the cost will be within the reach of industrial plants. What will be considered a feasible cost of control will depend on the effects of the pollutant on human health and the environment in general; indirectly, therefore, environmental quality is taken into consideration.

Emission/effluent standards are often used to define the "best practicable means" for single sources. Emission standards based upon "best practicable means" should in principle be uniform for the whole country, if the technical and economic conditions are essentially the same. With uniform emission standards, however, the environmental quality will vary with such factors as density of population and sources of environmental pollution, as well as with meteorological and hydrological conditions. Therefore, it seems reasonable that these be taken into account when choosing the "best practicable means" to be enforced. An advantage of the "best practicable means" approach is that different plants within an industry are treated alike so that no one company has an undue advantage. In addition, it serves to preserve environmental quality in non-polluted areas, since there is no incentive for industries to relocate to clean areas.

The disadvantages of this approach are that it does not guarantee a satisfactory environmental quality and that it does not greatly encourage the development of new or modified processes that produce less pollution or new abatement technology. Another criticism of the "best practicable means" approach is that in certain cases it may constitute control for control's sake, since it is not related to actual health effects.

Medium-related measures

Environmental quality or ambient standards

Environmental quality may be expressed in terms of one or more of a number of aspects. Among these are included health (of human, aquatic, and/or wildlife populations), economic considerations (for example, the environment has an economic quality if it facilitates the provision of agriculture and/or industrial goods and services), and recreational and aesthetic factors. At present, human health and economic considerations are viewed as the most important of these by society.

An environmental quality standard designed to attain or maintain the human "health" quality is a measurable concentration of a contaminant or other parameter of human stress, derived from known relationships between it and its effects. It may be expressed in terms of the mean of a number of samples collected during one or more periods of time, together with a maximum allowable deviation of an individual sample from the mean that will confer a predicted degree of health protection for the human population(s) involved.

An environmental quality standard may range from (1) a stated concentration value for a contaminant or other parameter of quality in an environmental medium designed to provide a degree of health or other protection for man or other species to (2) a regulatory standard, complete with statements of scope and purpose, definitions, specification of permissible concentration for the contaminant, compliance requirements, prescription of sampling methods and frequency, and acceptable analytical methods.

Environmental quality standards are often related to emission standards. In the United Kingdom, emission standards are set individually to achieve standards of water quality that are determined by the intended use of the water. The emission standards may thus be tightened if the monitoring of water content shows that quality levels are not being attained.

Exposure standards

Exposure standards are commonly set for the workplace and for food, but are also used for drinking-water quality control. The acceptable daily intake (ADI) (see pp. 20–23) is the most usual form of exposure standard; it specifies a daily level of chemical intake that is considered safe over a lifetime of exposure.

Exposure levels have been set for cadmium in drinking-water and bottled water in some countries. In 1972, an FAO/WHO Joint Expert Committee on Food Additives (see Annex 5) set a provisional maximum tolerable weekly intake from all sources of 400–500 µg of cadmium. Acceptable levels of exposure via the air have also been set for the workplace in many countries.

Biological standards

These are standards that relate to the concentration of pollutants in biological fluids and tissues, for example, lead in blood or mercury in human hair. The advantage of biological standards is that they include contamination from all sources, which may vary from person to person, thus their use provides an accurate picture of exposure. One disadvantage is that it is difficult to implement biological standards because this may require compulsory sampling. However, such standards can be useful for screening on the basis of volunteer samples. For example, the European Economic Community set three reference blood levels of lead for a screening survey. After this survey, in the United Kingdom, it was decided that when a child was found to have a blood level of more than 250 µg of lead per litre, his or her environment should be investigated for lead contamination (Haig, 1984a, 1984b).

Environmental quality management

This approach is based on environmental quality standards. From emission inventories and monitoring data, a dispersion model for the area concerned can be developed. Using this model, it is possible to calculate the reductions in emissions/effluents necessary to meet the standards adopted.

Environmental quality management is a logical approach. It permits long-term planning to avoid environmental pollution. It is possible to incorporate various factors such as city planning and energy and transport policies into programmes designed to achieve or maintain desired environmental quality. The approach has a built-in alert system to prevent environmental pollution from reaching concentrations that are in excess of those stipulated by the environmental quality standard.

One of the drawbacks of this approach is that there are serious gaps in the techniques for making emission inventories and developing dispersion models. There is also a need for continuous reappraisal of the adequacy of the predictions made in the planning stages as additional information becomes available on the performance of the controls.

Chapter 6

Legal framework

Environmental control policies define not only the responsibilities of industry, but also the limits on the government's power to intervene. The decision to undertake or forego regulation is usually controlled by an evaluation of risk to establish the circumstances justifying intervention. For example, regulatory action may be authorized whenever there is a substantial hazard to human life or health or the environment. In some instances a broader approach may be adopted, that requires the implementing agencies to weigh economic and social considerations together with public health aspects in reaching a decision concerning regulation.

Nature of the decision

The control options discussed in this book all seek to achieve one final result: to ensure that levels of pollutants in a given medium—air, water, soil, food—do not exceed specified quantities. In general, this may be achieved through either legislation or voluntary cooperation between industry and government, together with public information and education.

These two contrasting approaches towards improving the quality of the physical environment and safeguarding the health of the population are well illustrated by the methods adopted in the USA and the United Kingdom. On the whole, environmental regulation in the United Kingdom is relatively informal and flexible while regulations in the USA tend to be more formal and fixed. The United Kingdom makes extensive use of self-regulation and encourages close cooperation between government officials and representatives of industry, while in the USA the legislature and courts play a more active role in making and enforcing environmental policy (Vogel, 1986; Brickman et al., 1985).

The advantages of the cooperative approach can include effective collaboration among all parties involved and a shared commitment to improve steadily the quality of the environment. One disadvantage is the lack of accountability. In this case, the success of the regulating procedure depends on the sincerity with which the process is conducted and whether independent analyses are available. In short, both the regulators and the regulated must be committed to achieving socially acceptable levels of protection against risk.

Generally speaking, the choice of measures and strategies for regulation depends to a large extent on the existing legal and social framework in the country (Magat, 1982). In countries where there is a history of close cooperation between government and industry and of industrial compliance with government "suggestions", and a positive public attitude towards environmental quality, standards will tend to be applied through voluntary agreements. On the other hand, countries that possess a long history of state control of polluting activities through licensing, and a well developed administrative apparatus for the supervision of such activities, will generally use regulatory options that fit into this particular system.

Legislation is particularly advantageous in countries where the national constitution has provisions regarding the priority to be accorded to environmental protection measures and the rights and duties of the citizens with regard to the environment in which they live and work. In this sense, environmental law should not be understood as just another system of rules and agencies (Mayda, 1985). Rather, it must be viewed as part of a comprehensive process of resource management, supported by scientific data and assessment, and integrated with economic development planning.

Scope and development of environmental legislation

While the enactment of legislation is a political process, in many countries statutes and regulations are often drafted initially by technical and legal experts within government departments. In countries where several statutes dealing with a particular issue have been passed over a period of time, two evolutionary trends can be seen: development of more comprehensive laws and an increasingly creative and forward-looking role of government in environmental management.

For example, pollution control in many countries has evolved from common law, dealing with nuisances, into legislative control of emissions (or effluents) of particular pollutants at specific locations (e.g., chimney stacks or river outflow pipes) on a case-by-case basis. Other legislation involves the setting of ambient air and water quality standards that then determine what emission concentrations are permissible. Current legislation often attempts to anticipate and prevent potential environmental problems, e.g., through environmental impact assessment.

Over the years, the legislative framework has moved from a responsive role that facilitated particular decisions towards a guiding role that places decisions regarding pollution within a wider context of social and economic development (Whyte & Burton, 1980). Legislation in many countries now covers:

—protection and conservation of natural resources essential for survival of people, animals, and vegetation alike;

—national and international rules and regulations controlling the discharge and distribution of pollutants and chemical substances;

—approaches to planning and other behavioural controls to reduce the impact of man-made pollution and natural hazards; and

—the establishment of investigatory bodies to appraise, review, and quantify risks in relation to associated benefits.

The constitutional framework of a country determines, to a great extent, the permissible areas of regulation and the assignment and delegation of executive and administrative powers and responsibilities. Even in countries that do not have a constitution, public health and social policy considerations normally provide sufficient justification for the framing of environmental regulatory measures, provided these come within the permissible limits, e.g., concerning division of power. Countries with a federal structure must recognize the need to divide both jurisdictional and operational power and responsibility between the central (or federal) government and provincial (or state) governments.

The development of a legal framework comprises two regulatory actions. The first is the enactment of a formal legal instrument, such as an act, ordinance or decree. The second is the development of regulations, by-laws, rules, and orders, which may embody subsidiary legal changes, by the authority designated in the formal legal instrument. The development of these subsidiary instruments involves the determination and acceptance of a standard or norm, representing the pollutant level that is considered acceptable for protection of human health and the environment. In common speech, however, the words "regulation" and "standard" are often used interchangeably.

There are three possible approaches to effecting legal changes (Jayasuriya, 1985):

(a) the revision or updating of existing laws by amending legislation;
(b) the replacement of existing laws by an entirely new piece of legislation; and
(c) the enactment of comprehensive legislation, where none existed previously, by the consolidation and revision of sections from several existing laws, supplemented by new or additional provisions.

Approaches (a) and (b) are closely interlinked. It is often difficult to decide which of these two approaches is preferable in a particular situation. An assessment of the nature and magnitude of the contemplated changes and the ease with which they can be accommodated within the existing legislative framework may provide an indication of which approach is preferable.

Approach (c) is generally more cumbersome and time-consuming, particularly if substantial changes have to be effected in a number of statutes. Moreover, since the new provisions will be embodied in several different statutory instruments, the process of identifying the changes and new legislative structures and linking up the different substantive and procedural provisions is likely to prove tedious. In addition, if the various statutes to be amended come within the scope of different authorities or agencies, the new provisions may not be enforced uniformly unless there is an effective coordinating and monitoring mechanism.

Content and structure of regulations

As explained in the previous section, the fundamental decisions relating to acceptable risk can be found in the appropriate legislation; the technical details are usually set out in subsidiary regulations. For example, the main legislative instrument might empower the issue of licences to construct, operate, and maintain a certain manufacturing facility, but the format of such licences and the general conditions subject to which such licences may be issued might be dealt with in subordinate legislation.

It is difficult to draw an appropriate dividing line between the content of parliamentary legislation and subsidiary regulations. In a number of countries, there are constitutional provisions that require legislation for certain fundamental matters. There is also a question of expediency. Legislation, as a rule, is more difficult to amend and to adapt to changing circumstances, or technical possibilities, than are regulations. The process of making regulations also facilitates the participation of public interest groups and associations with expertise in the particular area concerned. Flexibility in the establishment of regulations is of particular importance in the field of environmental control (Bothe, 1980).

As a rule, the determination of quantitative ambient or emission standards and the prescription of certain known technical procedures for processes and use standards are relatively easy to handle. However, there are two problems. Firstly, a pollutant may be reaching the target through various pathways, which may vary in importance according to local conditions; in terms of health goals, therefore, it is difficult to define absolute acceptable concentrations of any specific pollutant in a given medium (Homburger, 1983). Quite often, this concentration can only be determined on a case-by-case basis. Secondly, new problems or dangers may emerge over time, and once recognized, may require that a stricter standard be set. A standard that has been determined on the basis of technical feasibility may well prove to be unnecessarily lenient or obsolete when new technologies are developed that require stricter standards (Stoel et al., 1980). In view of these two problems, regulations must be flexible enough to permit amendments to be made easily as new information and techniques become available.

Standards are generally based on evaluations of scientific data and information and seek to apply generalized risk assessment to individual situations. This, in turn, brings about the use of general clauses in the relevant regulation, which may be based on one of two different approaches. The regulation may take as a point of departure the danger to a specific target. Then, any pollution that constitutes an unacceptable danger to that particular target will be illegal. The alternative point of departure is technical feasibility. In other words, any pollution that can be avoided is illegal, even if it has not been proven that this pollution is harmful to a particular target. A number of countries have followed the latter approach in drawing up regulations. In this case, the standards used are related in some way to what may be called the "state of the art". This may be framed in terms of the best technology currently in use in a particular industry, the best technology available, e.g., in pilot projects and the like, or technology that is not yet developed but which will have to be developed if the potential polluter wants to start or carry on his activities.

The reference to the state of the art (however defined) is often combined with considerations of practical feasibility and economic costs. The legal problem related to the underlying economic issue is proportionality, a principle that, in different forms under different constitutional systems, limits state interference with private activity, in particular, economic activity. Thus, if the advantages for the environment gained by the use of a particularly expensive pollution abatement device are minimal, it may be a violation of this principle of proportionality to impose the use of such a device on a particular polluter.

A particular case in point concerns the "no-risk" or zero exposure principle. The guiding philosophy of this principle is that there is no acceptable level of risk for the hazard in question. This is in direct contrast to standards based on the threshold level approach, which inherently incorporates a risk–benefit analysis. Zero exposure puts the emphasis on the "risk" side of the equation.

Zero exposure is required most often when a country is considering the risks from suspected carcinogens. Experience in the USA with zero tolerances in legislation has been widely discussed. The best known legislation is the Delaney clause which regulates additives in food. In the United Kingdom, the zero exposure principle is the basis of the Pesticides Safety Precaution Scheme under which British companies may not sell pesticides that have not been cleared as non-carcinogens.

One difficulty with using a zero-exposure or zero-tolerance approach in standards is that the definition of "zero" depends on the accuracy of the measuring techniques available. Moreover, our present ability to detect amounts as low as one part per trillion for some substances far outstrips our ability to evaluate the significance of such minute amounts.

The format and structure of standards incorporated in legal instruments vary from country to country. However, any regulation will generally contain the following:

(*a*) a statement of the scope of the legal instrument, together with a definition of the applicable terms and concepts, specification of the responsible authorities, and of the parties, areas and substances to whom or to which the instrument will apply;

(*b*) mention of applicable documents, such as other standards, specifications, and regulations;

(*c*) a detailed description of the requirements, including limits on pollutants, applicable tests, mandatory control methods, reporting requirements, etc. Where the requirements are to be implemented over a period of time, a timetable will be included;

(*d*) a specific statement of the monitoring, reporting, and inspection systems; and

(*e*) a statement describing applicable penalties for contraventions.

Some legal instruments may be more inclusive, covering such matters as the operative time period of the instrument and procedures for review. Where tax incentives or other fiscal benefits are available, they may be set out in the instrument itself or in other relevant instruments such as the legislation concerned with taxation.

The authority empowered to enact regulations varies from country to country. As regards environmental standards, the appropriate authority may be the Minister of Health or the Environment Agency or Control Board. In many legal systems, however, the parliament or similar law-making body has the power to review and approve such instruments. The requirement that these instruments be published, usually in a government publication, and that they be tabled in the legislature, enables control to be exercised over their scope and content.

Experience from most parts of the world suggests that approval of standards is often nothing more than a procedural formality. Furthermore, certain legal systems do not require all standards and their revisions to be tabled in the legislature or other appropriate forum. For instance, the power to modify lists of hazardous substances or of scheduled industries can usually be exercised through an order that requires hardly any follow-up formalities, except perhaps its publication.

Institutional considerations

Different countries have different political and administrative hierarchies. Decisions, both legislative and administrative, are generally made at various levels, e.g., national, regional, and local. The distribution of power within the country determines the level at which particular decisions may be taken. In view of the complexity of many environmental issues, some thought must be given to the political and administrative organs and structures best suited to environmental decision-making in each country. Administrative considerations also influence the types of institutional mechanism and appellate procedure that are to be built into the legislation. For instance, the right to a hearing, before an application is rejected or refused or a licence is cancelled or suspended, has been incorporated into many administrative law systems.

One factor in determining the appropriate decision-making level for environmental control is the geographical scope of the problem. Another is the technical expertise of the various decision-making bodies. Authorities at the local level may not always possess the necessary knowledge to make decisions that take into account fully all the technical issues, possibilities, and options involved. They may also be more subject to local pressure than a central or national authority. On the other hand, the local conditions of the area affected must also be considered in the decision-making process.

The decision regarding the appropriate level is also related to whether the contemplated regulations are to be local in operation or applied on a nationwide basis. In some cases uniformity is necessary, in particular where the environmental problem is widespread or potential environmental regulations may create barriers to trade. Thus, pollution created by mobile sources is often regulated on a nationwide basis and, at least as far as certain basic minimum rules are concerned, regulations generally extend to the entire area in which these sources move; for example, pollution from aircraft is regulated on a worldwide basis. On the other hand, in situations where regional or local conditions warrant special consideration, the adaptation of general regulations may be inappropriate.

Another issue that should be considered in the drawing up of laws governing environmental pollution is that environmental risks are characteristically multidimensional problems that cut across the normal jurisdictions of government departments. Decisions that affect the environment may have to be made by public bodies or agencies that are not primarily concerned with environmental matters. One way of ensuring that environmental concerns are included in the decision-making process is to institute procedures providing for the participation of the competent government agencies in any decision that may have an impact on the environment. These procedures may take different forms, such as providing for representation in drafting or planning committees, or requiring submission of drafts for prior approval.

Where different agencies attempt to solve one particular problem together, several administrative problems may arise (Whyte & Burton, 1980):

—There may be uncertainty about exactly which agency should take overall responsibility and, as a result, action is delayed or not taken at all.

—Interdepartmental rivalries and jealousies may result in the withholding of information that should be shared, in order for the best solutions to be found.

—Each agency tends to have its own particular interests and constituency so that interagency conflict rather than cooperative problem-solving may ensue.

—Technical expertise may be divided between different agencies so that no one agency can gather together the necessary scientific and managerial team.

Countries that have special ministries, departments, or agencies with responsibility for environmental issues should ensure that all relevant sectors are adequately represented and that all perspectives are taken into account during policy- and decision-making (MacAndrews & Chia, 1979). Environmental issues cannot be treated in isolation; every effort must be made to take account of all factors that impinge on environmental problems.

Geographical dimension of environmental problems

The geographical range of environmental problems is essentially determined by the laws of nature. The spread of pollution in the air is governed by natural agents such as the wind, temperature distribution, and humidity. The spread of pollutants in water is determined by the natural flow of that water. Any meaningful regulatory approach must take these factors into account (Organization for Economic Cooperation and Development, 1984). Certain pollutants may find their way from one medium to another, e.g., from the air into soil and water and thus into the food chain. Certain areas are particularly vulnerable; others are particularly valuable because of their influence on climate, their capacity to produce oxygen, etc.

Natural phenomena, such as those described above, have nothing to do with the distribution of powers provided for by particular national legal systems or political boundaries drawn up by man. Thus, the jurisdictional scope of an agency competent to make the relevant decisions may not have the same boundaries as the geographical range of an environmental problem. This can pose problems within a country and even greater problems when environmental pollution transcends national boundaries. Generally, it is not possible to change the distribution of power in order to extend the territorial jurisdiction of an agency. In a federal country such a change of distribution of powers may necessitate constitutional amendments, which are difficult, if not impossible, to carry through. The solution to this problem, both within countries and especially in the international sphere, depends on the establishment of procedures of cooperation to ensure that the problem is dealt with in a consistent and concerted way, after consultation and deliberation with all relevant agencies, bodies or institutions.

Supporting measures for regulatory strategies

Where regulatory strategies have been adopted, the crucial question is whether the relevant standards are observed in practice. Various schemes and procedures, usually with a legislative basis, have been developed to facilitate the implementation of environmental standards.

Environmental impact assessment

An environmental impact assessment (EIA) is an assessment of the short-term and long-term effects of any proposed action (or absence of action), including policies, legislative proposals, programmes, projects, and operational practices, on the population and the physical, biological, and socioeconomic environment. It employs a structured approach to the evaluation of the impact of a development (e.g., hydroelectric installation, nuclear power plant, industrial complex), programme or policy on human health and the environment. Possible measures to prevent or control the impact of the various associated forms of environmental pollution are also included for consideration during the decision-making process. This allows decision-makers to select the option that results in the least pollution, or to stop the project entirely. Where standards exist, the persons proposing the development, programme or policy are required to state how they intend to comply with such standards. Environmental impact assessment is probably the most comprehensive procedure available today in which health effects and control strategies are considered (Mayda, 1985).

Licensing strategy

Where a licence is required before a product can be marketed, an installation built or operated, or a process used, but no rules are prescribed as to the particular characteristics that the product, installation, or process should possess, the licensing requirement is itself a strategy. On the other hand, if licensing is used as a way to ensure compliance with predetermined standards, it becomes an instrument for the enforcement of those standards.

The licensing requirement ensures that the proposed activities are not undertaken before the competent authority has approved the project on the basis of the information provided in the application. Licensing can also be used as a strategy that combines information gathering and licensing. Thus, a licence may be considered to have been granted if the competent authority does not react to information given within a prescribed period of time. Non-compliance with the conditions specified in the licence generally entails withdrawal, cancellation or suspension of the licence by the issuing authority. The possibility of such a course of action, which could have far-reaching economic implications, often promotes compliance with standards and other relevant conditions.

Dissemination of information

This aspect of regulation implementation is related to a more general subject—the shaping of public attitudes. Such attitudes may be influenced by, among other things, information provided by the state, or by private concerns or individuals in compliance with legislation. One particular aspect of this subject that is relevant to the control options discussed earlier is the attitude of consumers to products that are created by polluters or potential polluters. Thus, the enactment of regulations requiring that product labels contain information about possible hazards associated with the particular product may deter people from buying or using it. In this case, industry may then limit its production or produce alternatives that are less hazardous. Conversely, special labels for goods that possess positive qualities for the environment may preferentially stimulate their sale to the detriment of other, more harmful, products. Legal provision for environmental education programmes is useful in sensitizing the target groups of such programmes to the need to protect the environment and natural resources, using a variety of methods, including enforcement of standards.

Economic incentives or disincentives

Fiscal instruments, such as taxes, effluent charges, charges on particular pollutants, subsidies for desirable activities, and tax-deduction schemes, are quite often used as part of environmental policy (International Union for Conservation of Nature and Natural Resources, 1980a). They may be related to a particular standard and used to enforce it; for example, a charge may have to be paid by the polluter if an emission exceeds a certain prescribed concentration level. At present, the principal instruments used are taxes or other charges on emissions or other discharges of pollutants into the environment; charges on particular uses or processes; product charges (e.g., charges on oil, beverage containers or used tyres); direct subsidies for pollution abatement measures or for the development of relevant technology; and increased depreciation or other tax-deduction schemes for desirable investment, such as in pollution abatement equipment.

Care must be taken to define the purpose of such charges. Some may simply be used as a means of financing a certain activity, in which case they constitute a device for the redistribution of income. Others may be used to implement environmental policy by financing measures for the preservation or amelioration of the environment. Charges may also be levied against polluters in an attempt to deter potential offenders. In this case, the effluent or emission charge is one of the most widely discussed tools of environmental policy. It is said to be more effective than direct regulation (i.e., setting of binding standards) because it relies on market forces for protection of natural resources (in particular the regenerative capacity of air and water). The disincentive effect of effluent charges is only in part a question of market mechanisms, in that the charge is not based on supply and demand, but is determined unilaterally by the government.

The basis of the calculation and the determination of the charge, however, are crucial. The disincentive effect depends to some extent on the economic status of the polluter. The charge levied against the polluter must be higher than the cost of equivalent pollution abatement. The charge should be calculated on the basis of the amount of pollutant emitted into the environment. This poses a difficult administrative problem as regards measurement or estimation of the level of pollution. This practical difficulty appears to be the major obstacle to the extensive use of pollution charges as part of environmental policy.

A positive incentive for pollution abatement may often be appropriate, although such an approach also presents some fundamental problems. It is generally agreed by most industrialized states that it is the polluter, and not the taxpayer, who should bear the cost of pollution abatement or prevention. The adoption of this "no-subsidy" principle internationally means that distortions to trade are minimized. However, it has always been accepted that there are exceptions to this principle, for example, where an industry providing essential employment would be priced out of the market if it had to bear the cost of controlling its effluents. If strict environmental standards are to be enforced, some kind of financial assistance or tax subsidy from the government will often be needed, at least provisionally, so that weaker enterprises are not forced out of business.

Government subsidies, either in the form of direct subsidies or tax advantages (tax credits or accelerated depreciation rates), raise certain questions. It is important that these advantages are not given to finance investments that would be made even without the external financial incentive. Such unwarranted financial advantages merely provide windfall profits and do not act as an incentive.

It is, however, difficult, if not impossible, to ensure that such fiscal advantages are given only for investments that otherwise would not be made. Furthermore, if the provision of subsidies or the contribution of financial incentives is made to decrease industry's costs of investment, the assumption is made that the polluter is prepared to spend his own resources on pollution control. Any subsidy related to capital investment stimulates capital-intensive abatement measures (acquisition of expensive equipment) rather than development of other ways of reducing pollution (for instance, changes in the production process, in the material used, or in the product). It seems advisable that assistance should not be based on the introduction of any particular investment or technique, but rather on the simple criterion of damage reduction.

In addition, tax advantages, such as accelerated depreciation allowances, tend to favour highly profitable undertakings (those paying high taxes), and not the weaker, marginal enterprises, which may need more assistance and which may in some cases be the worst polluters.

Information gathering

Legal regulations for the protection of the environment often prescribe information that must be provided to a particular government body, usually a state agency. Such information may be necessary in relation to certain control strategies (for example, to ensure that emission standards are complied with), but information gathering and inspection may represent a regulatory strategy in itself. The need to give information, for example, on the possible effects of a product or on the possible impact of an installation, has the immediate effect of inducing the providers of such information to consider the potential consequences. In addition, this information provides the basis for any necessary feedback and follow-up action by the responsible agency. In the case of chemicals, a number of laws require the testing of a new chemical before it can be marketed, without defining any particular standard. Thus in this case, the obligation to provide information does not serve as a means of enforcing a particular standard, but rather as the basis for future action, if the relevant administration believes that marketing of the substance would be dangerous or would create an unacceptable risk (Brickman et al., 1985).

Administrative supervision and penalties

Preventive control through licensing requirements is not appropriate for all activities that may affect the environment. Even where they are used, licensing requirements have to be supplemented by regular checks. The supervising administration must not only possess the necessary information about what is going on, but must also be in a position to take effective measures on the basis of such information. Monitoring of ambient quality, as well as of the emissions or discharges of individual polluters, is essential. Furthermore, rules facilitating access by the competent agencies to information generated by the polluter are an important feature of effective supervision. For example, polluters may be obliged to measure emissions or discharges on a regular basis and to report the results. The agency will also have the right to carry out inspections, and to frame requirements for specific control measures, if, for example, a particular production process proves to be too harmful to the environment.

Compliance with applicable standards may also be ensured by administrative penalties or even criminal sanctions. One of the problems in this context is similar to that discussed in relation to fiscal disincentives, i.e., the penalties must be more costly than pollution abatement. Otherwise, the polluter will rather pay the penalties and continue to pollute. A higher penalty for a second or subsequent offence, including perhaps an order for closure or similar action, may improve compliance.

The mere existence of these measures of administrative supervision and sanctions may serve as a deterrent to potential polluters. In general, however, there are a certain number of conditions that must be met if the measures are to be effective. It has already been mentioned that sanctions must be sufficiently severe. Similarly, the supervisory or prosecuting agencies must be competent to deal with the problem; in particular, they must possess the necessary technical expertise and manpower, and they must have a certain amount of political backing. Polluters may try to exert political or other pressures on the agencies in an attempt to persuade them not to enforce the relevant standards or to interpret the regulations more leniently. Compliance with the applicable rules of environmental law must be accorded the highest priority, and safeguards must be taken against attempted interference with the exercise of powers under the law.

It is stressed that rules must be designed to ensure that the monitoring of compliance does not become too complicated and cumbersome. Since there are numerous activities that may have detrimental effects on the environment, regulatory bodies must focus on the most important aspects. Regulatory interventions should be introduced only where the chain of events allows for successful control. For instance, it is not feasible to monitor the emissions of individual motor vehicles; it is easier to formulate and enforce regulations on the construction of engines that will result in an overall reduction in exhaust emissions, although, in any particular case, emissions may be excessive due to a mechanical defect in an individual vehicle. A scheme for the periodic inspection of cars is mandatory. Subject to cost–benefit considerations, the feasibility of requiring owners of motor vehicles to install devices for the reduction of exhaust emissions may also be considered as a possible option. Such an option assumes importance particularly in countries that have a large number of old motor vehicles. Researchers and manufacturers need to be offered incentives to develop and manufacture such devices.

Land-use planning

Land-use planning is an important collateral method of protecting certain targets from pollution. By means of land-use planning, it is possible to prohibit environmentally harmful activities in certain areas and to confine polluting activities to clearly demarcated geographical areas or sites where no particularly vulnerable targets exist. Land-use planning is thus an important means of singling out and protecting areas that are vulnerable or particularly important from an ecological point of view.[1] If polluting activities are systematically confined to particular regions, the state becomes a planning agency for industrial siting decisions. Traditionally, it has been industry that has chosen suitable sites, leaving the state with only retrospective control over the consequences of a decision that has already been made (i.e., restrictive

---

[1] LAUSCHE, J. D. *Selected management tools for integrating environmental and health concerns with development*. Unpublished WHO document, EFP/EC/WP/83.13.

planning). In a situation where the capacity of the environment to eliminate pollution is limited, it may be more appropriate for the state itself to identify and demarcate appropriate sites and to exclude industrial installations from others (i.e., positive planning).

National conservation strategies

In 1980, the International Union for the Conservation of Nature and Natural Resources (IUCN) launched the World Conservation Strategy (International Union for Conservation of Nature and Natural Resources, 1980). The strategy has been widely endorsed by international organizations, governments, and non-governmental organizations throughout the world.

The strategy is now being used as a basis for discussion, programming, and action in a number of countries concerned with the ever-increasing pressures on their environmental resources and with the consequences of these pressures for long-term economic development (International Union for Conservation of Nature and Natural Resources, 1983). Some countries have developed National Conservation Strategies to emphasize the relationship between development goals that satisfy human needs and those that maintain a healthy environment. These strategies can form a vital part of the development planning process of a country; they can indicate patterns of development that are detrimental to the environment and help focus policies on alternative development schemes that satisfy human requirements while maintaining a healthy environment.

Chapter 7

Consequences of different approaches to environmental health protection

For each level of risk and for each means of control or prevention, there will be not only health-related concerns, but also a range of economic and social consequences. While an understanding of these consequences will obviously be very helpful in formulating a decision on control, it will not be necessary to apply all of the techniques described in this chapter to each situation. Rather, the analyses used should be designed to provide useful information on how effectively a proposal will achieve the desired result and how the proposal will affect other sectors of the society, such as the enforcing agency, the enterprises involved, and consumers. A full list of possible analyses is given in Table 5.

Table 5. Possible types of analysis of proposed control measures

Concern	Type of analysis
Risk	Net risk reduction Comparison of risks
Costs	Cost-effectiveness Industrial impact Economic impact Risk–cost comparisons
Benefits (and costs)	Benefit–cost analysis Distributional analysis
Environment, etc.	Cross-media analysis Environmental impact Social impact Other impacts

Risks and benefits

The first question to consider is whether the technical analyses have, in fact, covered all the risks and benefits associated with the proposed action. Often a decision to deal with one type of risk will affect other risks in society as well. These effects may be either positive (other risks are reduced) or negative (other risks are increased). For instance, prohibiting the use of nitrates as a food preservative, because the body can convert them into possible carcinogens, may actually increase the total health risk because nitrates in food prevent the development of *Clostridium botulinum*, a bacterium that causes food poisoning that is often fatal (World Health Organization, 1978a).

Such problems are commonly encountered by regulatory agencies in considering whether to limit the use of particular chemical substances. Usually, the imposition of such a limitation will result in the substitution of another substance for the one that is regulated. It then becomes necessary to consider the risks associated with the use, manufacture, and disposal of the substitute substance. Some possible substitutes may be more dangerous than the one being considered for control. It is necessary to determine whether the production process is likely to be more (or less) risky when these substitutions are made, and whether there will be more (or less) risk associated with the use of the product incorporating the substitute.

An industrial chemist familiar with the industrial processes under investigation may be able to answer many of these questions. Experience in countries that have already implemented the proposed controls may also provide useful information. Regulatory agencies in such countries may also be able to indicate what types of action can be taken to reduce any associated risks.

Cross-media concerns

It is also important to be sure that the proposed control measures will really reduce the risks and not just transfer them somewhere else (Conservation Foundation, 1984; World Resources Institute and International Institute for Environment and Development, 1986). For instance, some procedures for removing potentially toxic volatile chemicals from wastewater result in the release of these chemicals into the air. If this release occurs in an area with a high population density, the total risk may actually be increased by the effort to remove the chemicals from the water. Similarly, placing toxic substances in landfills may result in their being carried into surface water or groundwater by rainfall.

The procedures for conducting cross-media analyses are not well developed. Such an analysis requires a good understanding of the physical, chemical, and biological properties of the chemicals involved and the control techniques being proposed, as well as substantial insight and imagination. However, just asking what happens to the pollutant when a certain control is adopted may well result in the most serious cross-media problems being identified before they occur.

Acceptable risks

Having established the net risk reduction associated with the proposed action, the question arises as to whether this is considered acceptable by society (National Academy of Sciences, 1975). The types of risk that are considered acceptable can vary substantially from one society to another. A society's tolerance for risk may also decrease as it develops economically. This suggests that not only present but probable future levels of acceptable risk should be considered when making a decision.

What the public perceives as a significant hazard may be very different from the objectively assessed risk. For example, when three groups of people (league of women voters, university students, and business and professional club members) were asked to rank 30 causes of death, nuclear power was placed in first position by both the league of women voters and the university students. However, actuarial studies of the same 30 causes of death place nuclear power in 20th place (Upton, 1982). When the public perception of risk is very much higher, or lower, than the actuarial risk, difficulty may be experienced in formulating standards. The dilemma arises in deciding how much weight to give to the public perception of risk and how it can be reconciled with the computed risk.

Although some researchers have attempted to make rigorous analyses of a society's willingness to accept different types of risk, there is no accepted methodology for such comparisons, and no country has adopted such risk comparisons as the basis for regulation of hazards. Thus, there is likely to be little more than intuition, supported by public comment, to guide the decision-maker in evaluating the socially acceptable level of risk.

Cost and effectiveness of control measures

Rarely are regulatory decisions based solely on consideration of level of risk and available control techniques. The cost of the proposed action is usually also taken into account. A complete cost estimate includes both the "investment" cost required initially to bring facilities into line with the proposed regulations, and the costs required to maintain compliance thereafter, including any increased manufacturing costs and reduced profits that may result. Cost estimates based on experience in developed countries may not accurately represent the costs in less developed countries.

Usually, in addition to knowing the absolute cost of the proposed control method, the decision-maker wants to be sure that it is cost-effective, i.e., that it represents the cheapest way of achieving a given goal (Thompson, 1980).

The first problem in analysing cost-effectiveness is to define properly the goal to be achieved. Often a goal can be defined in different ways, which can have significant implications for the strategies to be considered and the resulting costs. For instance, the goal of a programme to control air pollution may be (a) to reduce air emissions to a given level, (b) to achieve a given ambient air quality, or (c) to reduce to an acceptable level the risks from the air pollutant. If goal (a) is adopted, the alternatives are limited to actions such as the installation of pollution abatement equipment to reduce the emissions to the desired level. If alternative (b) is chosen, the cost-effectiveness analysis would include all these techniques, as well as other strategies such as construction of higher smoke stacks or location of new facilities in clean areas. If goal (c) is chosen, the available alternatives may be expanded still further to include such items as improvement in medical care, location of facilities in remote areas where few people will be exposed to the pollutant, and requirements for workers to wear safety equipment.

The identification of all the alternatives is usually a substantial task. Once they have been identified, it usually becomes clear that they will not all, in fact, achieve the same goal. Commonly, the less expensive options will be less effective in reducing risk. For instance, the cheapest way of reducing the risk of hearing impairment in factories is usually for everyone working in the facility to wear ear-plugs. However, the employees often will not comply with such a regulation. Thus, while this alternative may appear to be the least expensive, it cannot be depended upon to achieve the desired reduction in risk.

The decision-maker, therefore, has to trade off the costs and effectiveness of the various possible control strategies. These trade-offs are further complicated by the probable sensitivity of the analyses to two sets of assumptions. The first set includes the assumptions made about the prices to be used in the cost estimates. Relatively small changes in interest rates or foreign exchange rates can produce large variations in the relative cost-effectiveness of alternative control options. The second set of assumptions includes those made about the relative prices, total emissions, and acceptable risks that will apply in the future. Changes in any of these may make it more cost-effective in the long-term to invest in more stringent controls than would seem to be necessary under existing conditions.

Other types of analysis

The types of analysis discussed so far, i.e., comparison of risks, analysis of acceptable risks, cross-media analyses, and cost-effectiveness analyses, should probably be included in some form in every decision-making process. Other types of impact analysis may be appropriate, depending on the particular circumstances and concerns.

Many impact analyses can be extremely expensive and time-consuming if strictly carried out. However, such strictness is often neither necessary nor useful. If a lack of resources, data, or valid methodology make rigorous analysis difficult or of doubtful value, there are several possible approaches. One is to make sure that someone familiar with the potentially important issues thinks carefully about how serious the impact of the proposed action may be. This, perhaps supplemented by a few discussions with other knowledgeable parties, may be all that is necessary. Such a quick review may suffice to eliminate concerns about the possible seriousness of the impact, or may allow the government or the affected parties to reduce the severity of any such impact.

A second possibility is to review studies relating to these issues that have been completed for other purposes. For instance, economic assessments undertaken to justify a foreign loan may contain enough information to suggest the likely economic impact of a pollution abatement requirement on a particular industry. Inferences can also be drawn from impact analyses carried out in other countries. Such inferences should only be drawn by someone famillar with the theory and methods used in the study, as well as with the particular types of problem that the proposed action may create.

A third approach is to conduct the entire decision-making process openly and publicly, inviting comments from individuals famillar with the issues raised.

All these actions may provide much of the information that would result from a more formal and rigorous analysis. They should at least help to ensure that the decision-maker is asking the right questions and examining the correct issues.

Risk–cost comparison

While cost-effectiveness analyses are used to compare alternative means of accomplishing the same reduction in risk for a particular pollutant, risk–cost comparisons are used to compare the relative costs of risk reduction, either by controlling a substance at different levels of stringency, or by reducing alternative types of risk.

Once estimates of both risk reduction and cost have been made, a risk–cost comparison is relatively straightforward. The reduction in risk (expressed in terms of the number of illnesses, accidents, or deaths avoided for the whole population) is divided into the cost of attaining that reduction to give a ratio measured in terms of the cost per unit risk reduction—for instance, dollars per life saved, per accident avoided, or per asthma attack avoided.

For such risk–cost analyses to be directly comparable with each other, the risk reduction and the costs must each be expressed in similar units. This requirement is not usually a problem as regards costs. It is, however, a problem when a proposed action will reduce more than one type of risk, or when the alternative actions affect different types of risk. In such cases, it is usually more helpful to carry out a benefit–cost analysis (see page 62).

Industrial impact analysis

An industrial impact analysis assesses the likely economic impact of the proposed regulations on the profits and viability of the industry most affected. These analyses are often inaccurate and actual compliance costs and economic impacts are frequently lower than projected. The industry's economic situation may also turn out to be much different from that expected. Any analysis should be reviewed by trained economists who are familar with the domestic and international performance of the industry.

A key determinant of the industrial impact is the extent to which industry can pass on any increased costs to the consumers in the form of higher prices. If, for instance, the industry is in close competition with foreign producers, it may be difficult to increase prices and the economic impact will therefore be significant. Another consideration is the size and age of individual companies in the industry: newer and larger companies are often better able to absorb increased costs.

An industrial impact analysis can help to identify actions that the government might take in order to reduce the severity of the economic impact. For instance, any economic impact will probably be reduced if the deadline for complying with the control requirements is delayed (but such delays may also extend the risks). Similarly, the controls could be applied immediately to new facilities, but be delayed for older ones. There are a number of such variations that can help reduce the economic impact and may be justified if the projected impact is unusually serious.

Economic impact analysis

Some countries undertaking extensive environmental and public health protection programmes have been concerned that the cost of these programmes might adversely affect the overall state of the nation's economy, particularly the rate of inflation, level of unemployment, economic growth, and foreign trade. Such large-scale effects can only be estimated by using relatively sophisticated macroeconomic models of the economy (Organization for Economic Cooperation and Development, 1985). The results of such efforts conducted in countries such as France, Japan, the Netherlands, and the USA generally show no significant impact. It is unlikely that similar analyses in developing countries would demonstrate significantly different results, unless the proposed expenditure was significantly greater than 2% of the country's gross domestic product. If such an analysis is thought to be desirable, the decision-maker should enlist the cooperation of the finance ministry or economic planning commission in conducting the analysis.

Benefit–cost analysis

When the effects of different types of action are being reviewed, both benefits and costs should generally be expressed in monetary units, and benefit–cost ratios computed in order to make inter-study comparisons easy (Organization for Economic Cooperation and Development, 1981; Thompson, 1980). Estimating these monetary values may be quite difficult, particularly when potentially fatal health risks or the stability of ecosystems is considered. Because of these problems, a sophisticated benefit–cost analysis may not always be a useful decision-making tool (Swartzman et al., 1982). However, even where economic values cannot be derived, the benefit–cost framework can assist in the organization of complex scientific data, by presenting meaningful quantitative measures of environmental results and providing a basis for comparison of improvements in health and the environment with the cost of pollution reduction (Luker, 1985).

Distributional analysis

Distributional analysis addresses such questions as whether any particular groups are expected to pay an unreasonable share of the costs of a proposed programme, and whether the benefits are widely distributed or limited to a small section of the population (Harrison, 1975). The distribution of the benefits of health and environmental protection programmes should be relatively clear from the risk assessment. It is also reasonably easy to identify those who are currently experiencing the risks and, therefore, will benefit when they are reduced.

Estimating the distribution of the costs, as well as any monetary benefits, may be more difficult. The costs will often be passed on through the economy so that they are ultimately met by people quite different from those who originally pay them. For instance, if the controls are imposed upon companies producing goods primarily for a protected domestic market in which there is little competition, the costs will probably be passed on to the consumer in the form of higher prices. On the other hand, an industry that exports much of its production and is in competition with other foreign companies (or with other substitute products) is essentially a "price taker" and will be unable to pass on the costs in the form of higher prices. In this case, the costs will be borne either by the owners of the company, by its workers (who may receive lower wages), or by the suppliers of the raw materials.

The question of the direction and the extent to which the costs are passed on requires analysis by a trained economist who is familar with the economy's characteristics and with the conditions in the industry facing the new expenditure. Because of the difficulty of making these analyses and the uncertainty that is necessarily associated with any results, distributional analysis might best be restricted to an analysis of the distribution of the direct benefits and the direct costs. Such a simplified analysis may well indicate whether or not there are likely to be any distributional problems associated with the proposed action.

Environmental impact analysis

An environmental impact analysis is often used to assess the environmental and social impacts of a proposed development project, such as a dam or a new power plant (Munn, 1979). Such analyses can also be undertaken for other types of action, such as the disposal of sludge from a sewage treatment plant or an air pollution scrubber in valuable wetland, or the construction of a sewer pipe which may damage valuable areas and interfere with natural stormwater run-off thus causing flooding. The purpose of an environmental impact statement is to identify both the possibility that undesirable events may occur and ways of avoiding them (see also page 51).

The decision as to which of these additional assessments should be undertaken for a given proposal will depend upon three considerations. The first is whether there are any particular national concerns (environmental, economic, or social) that may be adversely affected by the proposed action. For instance, certain environmental assets (e.g., wetlands, forests), certain "infant industries", or certain social groups (e.g., tribal people) may be considered particularly important. If any of these could be adversely affected by a proposed action, an impact statement may be called for. The second is whether any particular types of analysis, such as environmental impact assessment, are required by law, regulation, or national policy. The third is whether the proposed action is likely to have a major impact on one particular industrial sector, social group, environmental amenity, or geographical area of the country. When the impacts are concentrated in this fashion, it may be useful to evaluate them to ensure that they are not excessive.

Chapter 8

Decision-making process

It is not possible to describe in detail the decision-making process concerning regulation of an environmental problem, since this may involve a large number of activities and considerations. Nevertheless, the process does not always need to be extremely complex. Very often, for example, there are useful precedents, related either to the environment or other areas, which can serve as examples.

The main objectives in all instances are:

—to agree on a goal (or norm) for control of a particular pollutant, that will provide adequate protection for health; and
—to establish a suitable regulatory strategy and legislative instrument to achieve this health goal.

The decision-making process varies from country to country, even though from a strictly technical viewpoint the health goals are often similar. Variations may occur, for example, in developing countries where there may be a larger group of sensitive persons, thus necessitating a more stringent health goal. There will be differences in the strategies used to achieve the goal, depending on numerous factors including social, economic, and institutional considerations. In developing countries, for example, it may be suitable to have some form of licensing scheme that can be implemented by relatively few, well trained staff. In addition, the time-frame for implementing a certain measure may vary; for example, a country may decide to deal with new industries first, giving notice to existing older industries that they will be required to comply with the regulations within a certain number of years. Examples of this are the various control measures for cars, which are usually announced years ahead of time to give the automobile industry time to comply. Finally, however, despite the large number of individual national approaches to control of a given pollutant, there are few major differences and countries should not have much difficulty in finding appropriate examples relevant to their situation in other areas.

Environmental decision-making is never carried out in isolation. In recent years, public awareness of environmental hazards has increased considerably. This trend tends to increase the complexity of the process at the national level since the number of groups involved in discussions (government departments, politicians, industrialists, journalists, consumer groups, etc.) can be formidable (Vaupel, 1982; Coppock, 1985). In addition, consideration must be given to international implications of action at the national level. There are questions concerning the common environment (air and water) of neighbouring countries, and regarding the import and export of products. Plants and vegetables are often affected in this way when limits are set for the pesticide residues on (or in) them. These actions may sometimes be perceived as, or even used as, trade barriers. International conventions or agreements may also affect the national decision-making process. For example, countries bordering on the Mediterranean have signed a treaty that obliges them to limit their effluent from land-based sources. Certain groups of countries may decide to take joint action—this was done by the European Economic Community in the case of cadmium (see page 33), and by the countries belonging to the United Nations Economic Commission for Europe, which decided to reduce sulfur dioxide emissions to 70% of the 1980 level by 1993. Finally, citizens in one country may not be willing to accept a higher risk (i.e., have a more lenient norm or regulation) than those in other countries and the decision-maker may also have to take this popular feeling into consideration.

The number of environmental issues in each country requiring attention and possibly regulation can be large. Such issues may suddenly become areas of public concern as a result of many different occurrences, such as an industrial accident, press reports about a mysterious illness, or pressure for action from well organized and informed citizen groups. In all countries, the large number of items on the environmental agenda tends to stretch resources, such that some form of international cooperation or consultation for exchange of information and results of analysis is almost mandatory. In addition, the scientific information base will change with time as more data become available. This may lead to lower thresholds for no-observed-adverse-effects, and correspondingly stricter standards in some countries. This particular situation may cause problems in developing countries, since it takes time and it is sometimes difficult to find out the reasons why changes are made.

The organizational and institutional structures for making environmental decisions will vary from country to country (Schaefer, 1981; Johnson & Johnson, 1977; Lam, 1982; International Institute for Environment and Development, 1981). In the past, ministries of health were mainly involved in decision-making when basic sanitary measures in air, water, and food were the prime concern. With the increasing contamination of the environment as a result of waste disposal by industry and the ever-increasing use of chemicals for a multitude of applications, including agriculture, more and more sectors became involved in formulating measures for protecting people and the environment in general. Today, many countries have consolidated the responsibility for protecting the environment by setting up a separate department or ministry. Such a department may have mainly a coordinating function or, at the other extreme, may be a fully fledged environmental protection agency.

The complexity of the decision-making process will depend on the organizational and institutional structure. The existence of a department of the environment will probably facilitate the process, since it can establish a permanent mechanism for gathering and processing the information required for regulating pollutants. It can also ensure that the administration of the legislation and regulations is carried out in a well organized manner. Under these circumstances, the role of the ministry of health would be mainly to identify hazards, derive acceptable norms for the protection of the population, and carry out monitoring and health surveillance activities to ensure that the measures taken are adequate. Through a department of the environment, consultation with other departments and interested groups can be streamlined and made efficient.

The alternative to this procedure might be for a sectoral department (health, water resources, public works, labour, industry, etc.) to formulate the necessary proposal for environmental legislation or regulations and see it through until its adoption. This procedure will probably be more cumbersome and prolonged, but may present an advantage if the regulation contemplated will affect one particular sector or group of individuals, such as agriculture or industry.

Interactions between science and policy

The role of science in public decision-making has increased in parallel with the involvement of governments in the management of scientific and technological change. Caught up in a spiral of promotion and control of new techniques, the governments of all industrializing societies have to make complex policy choices requiring the regular input of information and advice from experts.

In order for an environmental protection programme to be successfully implemented, compromises will undoubtedly have to be accepted and improvements achieved will be short of the goals identified in studies of health effects and related analyses. The question arises, therefore, as to how important are such objective analyses. How far will they affect the political judgements made in relation to the proposed initiatives?

Drawing upon an empirical investigation of the way science is used in the regulation of toxic chemicals, a study in four industrialized countries (France, Federal Republic of Germany, the United Kingdom, and the USA) has shown that there is considerable variation in the extent to which governments rely on expert analysis and in how they acquire it (Brickman et al., 1985). While each country calls upon experts to identify and measure hazards and to clarify the available control options, the institutional means for incorporating these inputs and the broader purposes served differ considerably, as outlined below.

1. The United States exhibits a much stronger demand for expert analysis of all kinds in its regulatory processes than do the European

countries. Extensive use is made of scientific arguments and formal analysis in the regulatory process. The agencies involved have large research and development capabilities of their own, they have sizeable resources in terms of the expertise of the staff, and they perform the major part of their chemical risk assessments themselves. An important complementary source of expertise is outside consultants who are hired under agency contract.

The European countries provide a sharp contrast to this situation. The staff of the agencies that issue regulations rarely have the analytical skills needed to determine scientifically the level of risk. Unlike in the USA, where rules are accompanied by elaborate analytical justification and scientific documentation, European agencies rarely divulge the scientific rationale behind their decisions.

2. The European countries place more emphasis than the USA on the expertise of pluralistic advisory committees that combine technical competence with the representation of partisan viewpoints.

Advisory committees are a standard feature of toxic chemical control programmes in all four countries. However, European committees tend to be quite different in composition, role, and function from their American counterparts. They typically include representatives from the regulated industry and other interests, and their policy influence is greater than any similar body advising authorities in the USA.

3. Regulatory officials in the USA and the Federal Republic of Germany make most extensive use of the independent bodies of scientific and professional opinion. The two most visible and versatile purveyors of official scientific advice in these two countries are the National Academy of Sciences (NAS) and the Deutsche Forschungsgemeinschaft (DFG). These two organizations, which have no real equivalent in France or the United Kingdom, are frequently consulted by regulatory agencies for their recommendations and analyses. Official deference to their opinions appears to be greater in the Federal Republic of Germany than in the USA.

The decision-maker may be assisted by having a summary of the relevant information and analyses. When the decision has been made, it is wise to record the rationale behind the choice of the option selected. This will facilitate any judicial or administrative reviews of the decision, and it will enable others to understand the basis for the decision. It is particularly important when reviews are being carried out in the light of new data and information that may or may not have an impact on the decision taken.

Constraints affecting developing countries

Although many people believe that environmental controls in industrializing nations must be geared to their level of development and affluence, there is no concrete evidence to suggest that a country will attract international investment simply by minimizing the strictness of its environmental regulations. Neither is there any evidence to suggest that an industrializing country will deter multinational corporations simply by requiring them to adhere to environmental codes similar to those in effect in developed countries (MacAndrews & Chia, 1979).

Indeed, loose environmental constraints can pose a threat to foreign industries because excessive pollution can provoke strong anti-government and anti-multinational corporation sentiment among the population. As many examples demonstrate (Leonard, 1985), pollution frequently becomes an important local political issue. Citizens often resent what they perceive as collusion between their government and foreign companies in undertakings that despoil the environment or threaten public health. Thus, industrial strategies based on attracting pollution-creating industries can seldom, if ever, be tenable in the long term.

There are no fundamental reasons why the process for making decisions on environmental standards for the protection of the health of the people in developing countries should be different from that in developed countries. However, there are some specific factors that have a bearing on the process and may influence it in a particular way (Mayda, 1985). Some of these are outlined below.

Problem assessment

1. There are difficulties in the timely and effective transfer of information on specific pollutants that will allow national scientists and others to make evaluations and appropriate recommendations regarding national environmental issues. International efforts (see Annexes 1–7) are intended to help overcome this problem by providing consolidated information for the most important pollutants and chemicals.

2. Most of the health-related research information on pollutants is produced in developed countries with temperate climates and is based on studies of people who do not suffer from the debilitating effects of climatic stress, malnutrition, diarrhoea, malaria or general extreme poverty, all of which can impair the natural defence mechanisms of the body. It may therefore be necessary, until more data from developing countries become available, to consider with extra care the safety factors to be used in standard-setting in these countries. Increasing the safety factor would go some way towards ensuring that the relatively large, sensitive subpopulation is protected.

3. It is well known that, in developing countries, very rapid industrialization and urbanization are taking place. Both may be proceeding with very limited planning and without appropriate safeguards, thus exposing large numbers of people to much higher levels of environmental pollution within a short space of time (World Health Organization, 1977, 1985a).

The response mechanism

1. In most developing countries there is inevitably competition for resources between the economic and health development interests. Most often, a balance will have to be found in the form of a compromise decision being made on the levels of pollutants allowed for in the regulations (Bowonder, 1981).

2. Competing, and sometimes conflicting, health values may come to the fore; for example, attempts to control disease and increase food supply require the widespread use of hazardous insecticides and other pesticides that may in turn contaminate air, water, and food.

3. There may be inadequate mechanisms for coordinating the division of labour required to carry out the decision-making process leading to adequate legislation and standards. Also of concern in this regard may be a lack of vertical coordination within sectors (Schaefer, 1981).

Characteristics of an effective standard-setting process

From the points discussed, it can be seen that an effective process should have the following characteristics.

—It should involve, at some stage, the major parties in the community, such as politicians, bankers, citizen groups, industrial leaders, and health officials. Their involvement will stimulate debate encompassing differing perspectives and values; it will generally lead to some compromises being made in both goals and methods but will ensure broad support in the society at large.

—It should provide a mechanism through which technical and policy analyses can be generated, distributed, and criticised.

—It should provide a mechanism whereby the results of analyses can be presented to the policy-makers and the other centres of interest in the society, to inform these groups of the costs, benefits, and impact of the proposals under discussion.

—It should provide a mechanism for conflicting interests to be heard and discussed in a controlled manner, so that divergent opinions in the society can be aired and, as far as possible, accommodated in the implementation of the proposal.

—It should provide a mechanism whereby the society can reach a decision and take useful action, even though such action may be less than what is "objectively" ideal.

References

AMES, B. N. ET AL. Carcinogens are mutagens: a simple test system combining liver homogenates for activation and bacteria for detection. *Proceedings of the National Academy of Sciences of the United States of America*, 70: 2281–2285 (1973).

ASHBY, J. The unique role of rodents in the detection of possible human carcinogens and mutagens. *Mutation research*, 115: 177–213 (1983).

BIRGE, W. J. Structure–activity relationships in environmental toxicology. Introduction. *Fundamental and applied toxicology*, 3: 341–342 (1983).

BOTHE, M., ED. *Trends in environmental policy and law*. Gland, Switzerland, International Union for Conservation of Nature and Natural Resources, 1980.

BOWONDER, B. Environmental risk assessment issues in the third world. *Technology forecasting and social change*, 19: 99–127 (1981).

BRICKMAN, R. ET AL. *Controlling chemicals, the politics of regulations in Europe and the United States*. Ithaca, NY, and London, Cornell University Press, 1985.

BUŠTUEVA, K. A. [Principles for hygienic standardization of air pollutants] In: BUŠTUEVA, K. A., ed. [*Guidelines for hygiene of urban air*.] Moscow, Medicina, 1976, pp. 66–81 (in Russian).

CALABRESE, E. J. *Methodological approaches to deriving environmental and occupational health standards*. New York and Toronto, John Wiley & Sons, 1978.

CONSERVATION FOUNDATION. *State of the environment, an assessment at mid-decade*. Washington, DC, The Conservation Foundation, 1984.

COPPOCK, R. Interaction between scientists and public officials: a comparison of the use of science in regulatory programmes in the United States and West Germany. *Policy sciences*, 18: 371–390 (1985).

COUNCIL ON ENVIRONMENTAL QUALITY. *Fluorocarbons and the environment*. (Report of Federal Task Force on Inadvertent Modification of the Stratosphere). Washington, DC, US Government Printing Office, 1975.

DAVOS, C. A. & NIENBERG, M. W. A framework for integrating the health concern into environmental decision making. *Journal of environmental management*, 11: 133–146 (1980).

EISENREICH, S. J. ET AL. *Assessment of airborne organic contaminants in the great Lakes ecosystem*. Minneapolis, MN, Report to the Science Advisory Board's Ecological and Geochemical Aspects Expert Committee of the International Joint Committee, University of Minnesota Environmental Engineering Program, 1980.

EL BATAWI, M. A. & GOELZER, B. I. F. Internationally recommended health-based occupational exposure limits: a programme in the World Health Organization. *Annals of American industrial hygiene*, 12: 49–57 (1985).

ENVIRONMENTAL PROTECTION AGENCY (1976). *Quality criteria for water*. Washington, DC, Government Printing Office, 1976.

ENVIRONMENTAL PROTECTION AGENCY (1977). *Pre-screening for environmental hazards: a system for selecting and prioritizing chemicals*. Washington, DC, EPA, 1977 (EPA 560/1-77-001).

ENVIRONMENTAL PROTECTION AGENCY (1984). *Extent of hazardous release and future funding needs, section 301 (a)(1)(c) study of the comprehensive environmental response, compensation, and liability act. Final report, table 2–1*.

Washington, DC, EPA, Office of Solid Waste and Emergency Response, 1984.

ENVIRONMENTAL PROTECTION AGENCY (1985). *Chemical emergency procedures program. Interim guidance.* (Draft) Washington, DC, EPA, 1985.

EUROPEAN ECONOMIC COMMUNITY (1982). Council directive on the major accident hazards of certain industrial activities. *Official journal of the European communities,* No. L.230, 1982 (82/501/EEC).

EUROPEAN ECONOMIC COMMUNITY (1983). Council directive on limit values and quality objectives for cadmium discharges. *Official journal of the European communities,* No. L.291, 1983 (83/513/EEC).

GKNT. *Problems of industrial toxicology.* Moscow, USSR, Centre of International Projects, USSR State Committe for Science and Technology, 1986.

HAIG, N. (1984a). *Coordination of standards for chemicals for environmental protection.* London, Institute for European Environmental Policy, 1984.

HAIG, N. (1984b). *EEC Environmental policy and Britain.* London, Environmental Data Services Ltd., 1984.

HARRISON, D. *Who pays for clean air?* Cambridge, MA, Ballinger Publishing Co., 1975.

HOLDGATE, M. W. *A perspective of environmental pollution.* Cambridge, England, Cambridge University Press, 1979.

HOMBURGER, F., ED. *Safety evaluation and regulation of chemicals.* Basel, Karger, 1983.

HUISINGH, D. & BAILEY, V., ED. *Making pollution prevention pay: ecology with economy as policy.* New York, Pergamon Press, 1982.

INTERNATIONAL INSTITUTE FOR ENVIRONMENT AND DEVELOPMENT. *Legal, regulatory, and institutional aspects of environmental and natural resources management in developing countries.* Washington, DC, US Agency for International Development, 1981.

INTERNATIONAL LABOUR OFFICE. *Occupational exposure limits for airborne toxic substances,* 2nd revised edition. Geneva, ILO, 1980 (Occupational Safety and Health Series, No. 37).

INTERNATIONAL UNION FOR CONSERVATION OF NATURE AND NATURAL RESOURCES (1980). *World conservation strategy.* Gland, Switzerland, IUCN, 1980.

INTERNATIONAL UNION FOR CONSERVATION OF NATURE AND NATURAL RESOURCES (1980a). *Trends in environmental policy and law.* Gland, Switzerland, IUCN, 1980.

INTERNATIONAL UNION FOR CONSERVATION OF NATURE AND NATURAL RESOURCES (1983). *National conservation strategies* (Report to development assistance agencies on progress and planning for sustainable development). Gland, Switzerland, IUCN, 1983.

IZMEROV, N. F. *Control of air pollution in the USSR.* Geneva, World Health Organization, 1973 (Public Health Paper, No. 54).

JAYASURIYA, D. C. *Regulation of pharmaceuticals in developing countries.* Geneva, World Health Organization, 1985.

JOHNSON, H. & JOHNSON, J. M. *Environmental policies in developing countries.* Berlin (West), Erich Schmidt Verlag, 1977.

KATSUMA, M., ED. *Minamata disease.* Japan, Kumamoto University, 1968.

KLINE, J. ET AL. Surveillance of spontaneous abortions, power in environmental monitoring. *American journal of epidemiology,* 106: 345–350 (1977).

KNEESE, A. V. & BOWER, B. T. *Environmental quality and residuals management.* Baltimore, MD, Johns Hopkins University Press, 1979.

LAM, K. *Environmental management in developing countries: some lessons from China and Hong Kong.* Hong Kong, The Chinese University of Hong Kong, 1982 (Occasional paper, No. 49).

LEONARD, H. J. Confronting industrial pollution in rapidly industrializing countries: myths, pitfalls, and opportunities. *Ecology law quarterly*, **12**: 779–816 (1985).

LUKER, R. A. The emerging role of benefit–cost analysis in the regulatory process at EPA. *Environmental health perspectives*, **62**: 373–379 (1985).

MACANDREWS, C. & CHIA, L. S., ED. *Developing economies and the environment—the South-East Asia experience.* Singapore, McGraw Hill, 1979.

MAGAT, W. A. *Reform of environmental regulation.* Cambridge, MA, Ballinger Publishing Company, 1982.

MALES, R. Risks, risk taking, and societal decisions. *Environment international*, 495–499 (1985).

MAYDA, J. Environmental legislation in developing countries: some parameters and constraints. *Ecology law quarterly*, **12**: 997–1204 (1985).

MUNN, R. E., ED. *Environmental impact assessment: principles and procedures*, 2nd ed. SCOPE Report 5. London and New York, John Wiley and Sons, 1979.

NATIONAL ACADEMY OF SCIENCES (1975). *Decision-making for regulatory chemicals in the environment.* Washington, DC, NAS, 1975.

NATIONAL ACADEMY OF SCIENCES (1977). *Drinking water and health*, Vol. 1. Washington, DC, NAS, 1977.

NATIONAL RESEARCH COUNCIL. *Identifying and estimating the genetic impact of chemical mutagens.* Washington, DC, National Academy Press, 1983.

NIKIFOROV, B. ET AL. [Method for calculating maximum permissible concentrations of air pollutants.] *Gigiena i sanitarija*, **10**: 56–61 (1979) (in Russian).

ORGANIZATION FOR ECONOMIC COOPERATION AND DEVELOPMENT (1981). *The costs and benefits of sulphur oxide control.* Paris, OECD, 1981.

ORGANIZATION FOR ECONOMIC COOPERATION AND DEVELOPMENT (1984). *Chemicals on which data are currently inadequate: selection criteria for health and environmental purposes.* Rome, Istituto Superiore di Sanita, and Berlin (West), Umwelt Bundesamt, 1984.

ORGANIZATION FOR ECONOMIC COOPERATION AND DEVELOPMENT (1984a). *Emission standards for major air pollutants from energy facilities in OECD member countries.* Paris, OECD, 1984.

ORGANIZATION FOR ECONOMIC COOPERATION AND DEVELOPMENT (1985). *The macroeconomic impact of environmental expenditure.* Paris, OECD, 1985.

PINIGAN, M. A. [Scientific basis for hygienic protection of ambient air.] In: [*Sanitary protection of urban air*], Moscow, 1976, pp. 15–47 (in Russian).

PINIGAN, M. A. & GRIGOREVSKÁYA, Z. P. [Methodological approaches for establishing time-related maximum permissible concentrations for ambient air pollutants with seasonal differentiation.] [*Hygienic aspects of environmental protection*], **6**: 64–68 (1978) (in Russian).

RICCI, P. F. & MOLTON, L. S. Regulating cancer risks. *Environmental science and technology*, **19**: 473–479 (1985).

RODRICKS, J. V. & TARDIFF, R. G., ED. *Assessment and management of chemical risks.* Washington, DC, American Chemical Society, 1984 (ACS symposium series, No. 239).

RYAZANOV, V. A. [Principles for hygienic standardization of air pollutants.] In: Ryazanov, V. A., ed. [*Guidelines for community hygiene*], Vol. 1. Moscow, Medgiz, 1961, p. 194 (in Russian).

SCHAEFER, M. *Intersectoral coordination and health in environmental management*. Geneva, World Health Organization, 1981 (Public Health Paper, No. 74).

STOEL, T. B. ET AL. *Fluorocarbon regulation*. Lexington, MA, Lexington Books, 1980.

SWARTZMAN, D. ET AL. *Cost–benefit analysis and environmental regulations: politics, ethics and methods*. Washington, DC, The Conservation Foundation, 1982.

TECHNICAL INFORMATION PROJECT. *Toxic substances: decisions and values* (proceedings of a conference on decision making). Washington, DC, Technical Information Project Inc., 1979.

THOMPSON, M. S. *Benefit–cost analysis for program evaluation*. London and Beverley Hills, CA, Sage Publications, 1980.

UNITED NATIONS ENVIRONMENT PROGRAMME (1982). *List of dangerous chemical substances and processes*. Nairobi, UNEP, 1982 (UNEP/GL.10.5 Add.3).

UPTON, A. C. The biological effects of low-level ionizing radiation. *Scientific American*, **2**: 29–37 (1982).

USA. *Code of Federal Regulations—Protection of Environment*, **40**, Part 165.7 to 165.9. Washington, DC, US Government Printing Office, 1984.

VAINIO, H. ET AL. Data on carcinogenicity of chemicals in the IARC Monographs programme. *Carcinogenesis*, **6**: 1653–1665 (1975).

VAUPEL, J. W. Truth and consequences: some roles for scientists and analysts in environmental decision making. In: *Reform of environmental legislation*. Cambridge, MA, Ballinger Publishing Co., 1982.

VETTORAZZI, G. *Handbook of international food regulatory toxicology. Vol. I: Evaluations*. New York/London, SP Medical and Scientific Books, 1980.

VETTORAZZI, G. & RADAELLI–BENVENUTI, B. M. *International regulatory aspects for chemicals. Vol. II: Toxicological data profiles*. Boca Raton, FL, CRC Press Inc., 1982.

VOGEL, D. *National styles of regulation in environmental policy in Great Britain and the United States*. Ithaca, NY, University Press, 1986.

VON MOLTKE, K. ET AL. *The regulation of existing chemicals in the European Community: possibilities for the development of a community strategy for the control of cadmium*. London/Paris/Bonn, Institute for European Environmental Policy, 1985 (Rept. 84-B6602-11-0006-11-N).

WHYTE, A. V. & BURTON, I., ED. *Environmental risk assessment*. SCOPE Report 15. New York and Toronto, John Wiley and Sons, 1980.

WORLD HEALTH ORGANIZATION (1972). Technical Report Series, No. 506, 1972 (*Air quality criteria and guides for urban air pollutants*: report of a WHO Expert Committee).

WORLD HEALTH ORGANIZATION (1974). Technical Report Series, No. 554, 1974 (*Health aspects of environmental pollution control: planning and implementation of national programmes*: report of a WHO Expert Committee).

WORLD HEALTH ORGANIZATION (1975). *Methods used in the USSR for establishing biologically safe levels of toxic substances. Papers presented at a meeting in Moscow, 12–19 December 1972*. Geneva, World Health Organization, 1975.

WORLD HEALTH ORGANIZATION (1976). *Mercury*. Geneva, World Health Organization, 1976 (Environmental Health Criteria, No. 1).

WORLD HEALTH ORGANIZATION (1977). Technical Report Series, No. 601, 1977

(*Methods used in establishing permissible levels in occupational exposure to harmful agents*: report of a WHO Expert Committee with the participation of ILO).

WORLD HEALTH ORGANIZATION (1978). *Principles and methods for evaluating the toxicity of chemicals. Part 1.* Geneva, World Health Organization, 1978 (Environmental Health Criteria, No. 6).

WORLD HEALTH ORGANIZATION (1978a). *Nitrates, nitrites, and N-nitroso compounds.* Geneva, World Health Organization, 1978 (Environmental Health Criteria, No. 5).

WORLD HEALTH ORGANIZATION (1980). Technical Report Series, No. 647, 1980 (*Recommended health-based limits in occupational exposure to heavy metals*: report of a WHO Study Group).

WORLD HEALTH ORGANIZATION (1981). Technical Report Series, No. 664, 1981 (*Recommended health-based limits in occupational exposure to selected organic solvents*: report of a WHO Study Group).

WORLD HEALTH ORGANIZATION (1982). Technical Report Series, No. 677, 1982 (*Recommended health-based limits in occupational exposure to pesticides*: report of a WHO Study Group).

WORLD HEALTH ORGANIZATION (1982a). *Risk assessment* (proceedings of a seminar). Copenhagen, World Health Organization, 1982.

WORLD HEALTH ORGANIZATION (1982b). *Rapid assessment of air, water and land pollution.* Geneva, World Health Organization, 1982 (WHO Offset Publication, No. 62).

WORLD HEALTH ORGANIZATION (1983). *Guidelines on studies in epidemiology.* Geneva, World Health Organization, 1983 (Environmental Health Criteria, No. 27).

WORLD HEALTH ORGANIZATION (1983a). Technical Report Series, No. 684, 1983 (*Recommended health-based occupational exposure limits for selected vegetable dusts*: report of a WHO Study Group).

WORLD HEALTH ORGANIZATION (1984). *Guidelines for drinking-water quality. Vol. 1. Recommendations.* Geneva, World Health Organization, 1984.

WORLD HEALTH ORGANIZATION (1984a). Technical Report Series, No. 707, 1984 (*Recommended health-based occupational exposure limits for respiratory irritants*: report of a WHO Study Group).

WORLD HEALTH ORGANIZATION (1985). *Guidelines for drinking-water quality. Vol. 2. Health criteria and other supporting information.* Geneva, World Health Organization, 1985.

WORLD HEALTH ORGANIZATION (1985a). Technical Report Series, No. 718, 1985 (*Environmental pollution control in relation to development*: report of a WHO Expert Committee).

WORLD HEALTH ORGANIZATION (1987). *Air quality guidelines for the European region.* Copenhagen, WHO Regional Office for Europe, 1987.

WORLD HEALTH ORGANIZATION (1986a). Technical Report Series, No. 734, 1986 (*Recommended health-based limits in occupational exposure to selected mineral dusts (silica, coal)*: report of a WHO Study Group).

WORLD HEALTH ORGANIZATION (1986b). Technical Report Series, No. 733, 1986 (*Evaluation of certain food additives and contaminants*: twenty-ninth report of the joint FAO/WHO Expert Committee on Food Additives).

WORLD RESOURCES INSTITUTE AND INTERNATIONAL INSTITUTE FOR ENVIRONMENT AND DEVELOPMENT. *World resources 1986, an assessment of the resource base that supports the global economy.* New York, Basic Books Inc., 1986.

YANYŠEVA, N. Ya. [Methodological approaches to the standardization of carcinogenic compounds in ambient air.] *Gigiena i sanitarija*, 1: 90–93 (1972) (in Russian).

ZAEVA, G. N. [General quantitative relationship of some toxicometric parameters of substances.] In: [*Toxicology of new industrial chemicals*], Vol. 6. Moscow, 1964, pp. 165–180 (in Russian).

WHO environmental health criteria documents (as of September 1986)

1. Mercury (1976)
2. Polychlorinated biphenyls and terphenyls (1976)
3. Lead (1977)
4. Oxides of nitrogen (1977)
5. Nitrates, nitrites, and N-nitroso compounds (1978)
6. Principles and methods for evaluating the toxicity of chemicals, Part 1 (1978)
7. Photochemical oxidants (1978)
8. Sulfur oxides and suspended particulate matter (1979)
9. DDT and its derivatives (1979)
10. Carbon disulfide (1979)
11. Mycotoxins (1979)
12. Noise (1980)
13. Carbon monoxide (1979)
14. Ultraviolet radiation (1979)
15. Tin and organotin compounds (1980)
16. Radiofrequency and microwaves (1981)
17. Manganese (1981)
18. Arsenic (1981)
19. Hydrogen sulfide (1981)
20. Selected petroleum products (1982)
21. Chlorine and hydrogen chloride (1982)
22. Ultrasound (1982)
23. Lasers and optical radiation (1982)
24. Titanium (1982)
25. Selected radionuclides (1983)
26. Styrene (1983)
27. Guidelines on studies in environmental epidemiology (1983)
28. Acrylonitrile (1983)
29. 2,4-Dichlorophenoxyacetic acid (2,4-D) (1984)
30. Principles for evaluating health risks to progeny associated with exposure to chemicals during pregnancy (1984)
31. Tetrachloroethylene (1984)
32. Methylene chloride (1984)
33. Epichlorohydrin (1984)
34. Chlordane (1984)
35. Extremely low frequency (ELF) fields (1984)
36. Fluorides and fluorine (1984)
37. Aquatic (marine and freshwater) biotoxins (1984)
38. Heptachlor (1984)
39. Paraquat and diquat (1984)
40. Endosulfan (1984)
41. Quintozene (1984)
42. Tecnazene (1984)
43. Chlordecone (1984)
44. Mirex (1984)
45. Camphechlor (1984)
46. Guidelines for the study of genetic effects in human populations (1985)
47. Summary report on the evaluation of short-term tests for carcinogens (collaborative study on *in vitro* tests) (1985)
48. Dimethyl sulfate (1985)

49. Acrylamide (1985)
50. Trichloroethylene (1985)
51. Guide to short-term tests for detecting mutagenic and carcinogenic chemicals (1985)
52. Toluene (1986)
53. Asbestos and other natural mineral fibres (1986)
54. Ammonia (1986)
55. Ethylene oxide (1985)
56. Propylene oxide (1985)
57. Principles of toxico-kinetic studies (1986)
58. Selenium (1986)
59. Principles for evaluating health risks from chemicals during infancy and early childhood: the need for a special approach (1986)
60. Principles and methods for the assessment of neurotoxicity associated with exposure to chemicals (1986)
61. Chromium (1986)
62. 1,2-Dichloroethane (1986)
63. Organophosphorus insecticides—a general introduction (1986)
64. Carbamate pesticides—a general introduction (1986)
65. Butanols—four isomers (1986)
66. Kelevan (1986)
67. Tetradifon (1986)
68. Hydrazine (1986)

Example from the file of the International Register of Potentially Toxic Chemicals[a]

IRPTC				RECOMMENDATIONS—LEGAL MECHANISMS
TG0350000 FENITROTHION				
AREA	TYPE	SUBJECT	DESCR-IPTOR	LEVELS, REMARKS AND REF-ERENCE
BRA	REG	FOOD FEED	AL AL	PLANT (SPECIFIED): 0.1–0.4 MG/KG (SAFETY INTERVAL: 7–30 DAYS); PASTURE: 0.5 MG/KG (INTERVAL BETWEEN APPLICATION AND GRAZING: 10 DAYS) EFFECTIVE DATE: ENTRY DATE IN IRPTC: AUG 1982
				SOURCE: CATALOGO DOS DEFENSIVOS AGRICOLAS, 2, 85, 1980
CSK	REG		CLASS	POISONOUS SUBSTANCE EFFECTIVE DATE: JUL 1967 ENTRY DATE IN IRPTC: MAY 1982
				SOURCE: SBIRKA ZAKONU CESKOSLOVENSKE SOCIALISTICKE REPUBLIKY/COLLECTION OF THE LAW OF CZECHOSLOVAK SOCIALIST REPUBLIC, 22, 217, -, 1967
CSK	REG	HUMAN FOOD	MRL	LIMIT OF RESIDUES PRESENT DUE TO PLANT PROTECTION: 0.5 MG/KG. EFFECTIVE DATE: OCT 1978 ENTRY DATE IN IRPTC: MAY 1982
				SOURCE: HYGIENICKE PREDPISY MINISTERSTVA ZDAVOTNICTVI CSR/HYGIENIC REGULATIONS OF MINISTRY OF HEALTH OF CSR, 43, -, 1978

Annex 2 (Contd.)

	REG	USE	AGRIC	PRMT		
CSK	REG	USE	AGRIC	PRMT	SUBSTANCE IS APPROVED AS PESTICIDE. SPECIFIC USES, LIMITATIONS AND SAFETY PRECAUTIONS ARE GIVEN SOURCE: SEZZAM POVOLENYCH PRIPRAVKU NA OCHRANU ROSTLIN/LIST OF PERMITTED CHEMICALS FOR PLANT PROTECTION, -, 1981	EFFECTIVE DATE: AUG 1981 ENTRY DATE IN IRPTC: MAY 1982
DEU	REG	FOOD		MRL	PLANT (SPECIFIED) 0.5 MG/KG. SOURCE: BUNDESGESETZBLATT, IS. 718, 729, 1978	EFFECTIVE DATE: 1 AUG 1978 ENTRY DATE IN IRPTC: MCH 1982
DEU	REG	LABEL PACK USE	— — OCC	RQR RQR RSTR	FOR LABEL AND PACKAGING SEE EEC (OJEC**, L. 360, 1, 1976). HANDLING OF SOME GROUPS OF CHEMICALS (INCL. CARCINOGENIC, MUTAGENIC, POISONOUS, EXPLOSIVE AND EASILY INFLAMMABLE COMPOUNDS) IS PROHIBITED OR RESTRICTED FOR PREADULTS AND PREGNANT OR NURSING WOMEN. SOURCE: BUNDESGESETZBLATT, IS. 2069, 2069, 1980	EFFECTIVE DATE: 1 OCT 1980 ENTRY DATE IN IRPTC: JUN 1982

a Explanatory notes
BRA—Brazil
CSK—Czechoslovakia
DEU—Federal Republic of Germany
REG—Regulation
OCC—Occupation hygiene
AL—Acceptable limit(s)

CLASS—Classification
MRL—Maximum restriction limit
PRMT—Permitted
RQR—Requirement(s)
RSTR—Restriction(s)
SOURCE—Information reference

Annex 3

IARC monographs on the evaluation of the carcinogenic risk of chemicals to man (as of September 1986)

1. Some inorganic substances, chlorinated hydrocarbons, aromatic amines, N-nitroso compounds, and natural products (1972)
2. Some inorganic and organometallic compounds (1973)
3. Certain polycyclic aromatic hydrocarbons and heterocyclic compounds (1973)
4. Some aromatic amines, hydrazine and related substances, N-nitroso compounds and miscellaneous alkylating agents (1974)
5. Some organochlorine pesticides (1974)
6. Sex hormones (1974)
7. Some anti-thyroid and related substances, nitrofurans and industrial chemicals (1974)
8. Some aromatic azo compounds (1975)
9. Some aziridines, N-, S- and O-mustards and selenium (1975)
10. Some naturally occurring substances (1976)
11. Cadmium, nickel, some epoxides, miscellaneous industrial chemicals and general considerations on volatile anaesthetics (1976)
12. Some carbamates, thiocarbamates and carbazides (1976)

13. Some miscellaneous pharmaceutical substances (1977)
14. Asbestos (1977)
15. Some fumigants, the herbicides 2,4-D and 2,4,5-T, chlorinated dibenzodioxins and miscellaneous industrial chemicals (1977)
16. Some aromatic amines and related nitro compounds—hair dyes, colouring agents and miscellaneous industrial chemicals (1978)
17. Some N-nitroso compounds (1978)
18. Polychlorinated biphenyls and polybrominated biphenyls (1978)
19. Some monomers, plastics and synthetic elastomers, and acrolein (1979)
20. Some halogenated hydrocarbons (1979)
21. Sex hormones (II) (1979)
22. Some non-nutritive sweetening agents (1980)
23. Some metals and metallic compounds (1980)
24. Some pharmaceutical drugs (1980)
25. Wood, leather and some associated industries (1981)
26. Some antineoplastic and immunosuppressive agents (1981)

27. Some aromatic amines, anthraquinones and nitroso compounds, and inorganic fluorides used in drinking-water and dental preparations (1982)
28. The rubber industry (1982)
29. Some industrial chemicals and dyestuffs (1982)
30. Miscellaneous pesticides (1983)
31. Some food additives, feed additives and naturally occurring substances (1983)
32. Polynuclear aromatic compounds, Part 1, Chemical, environmental and experimental data (1983)
33. Polynuclear aromatic compounds, Part 2, Carbon blacks, mineral oils and some nitroarenes (1984)
34. Polynuclear aromatic compounds, Part 3, Some complex industrial exposures in aluminium production, coal gasification, coke production, and iron and steel founding (1984)
35. Polynuclear aromatic compounds, Part 4, Bitumens, coal-tars and derived products, shale-oils and soots (1985)
36. Some allyl compounds, aldehydes, epoxides and peroxides (1985)
37. Tobacco habits other than smoking, betel-quid and areca-nut chewing and some related nitrosamines (1985)
38. Tobacco smoking (1986)

Supplements

1. Chemicals and industrial processes associated with cancer in humans (IARC Monographs, Volumes 1–20) (1979)
2. Long-term and short-term screening assays for carcinogens: a critical appraisal (1980)
3. Cross index of synonyms and trade names in Volumes 1 to 26 (1982)
4. Chemicals, industrial processes and industries associated with cancer in humans (IARC Monographs, Volumes 1–29) (1982)

Pesticides for which evaluations of residues in food have been carried out (as of September 1986)

Further information regarding evaluation of specific pesticide residues is available from Division of Environmental Health, World Health Organization, 1211 Geneva 27, Switzerland.

Acephate
Acrylonitrile
Aldicarb
Aldrin
Aminocarb
Allethrin
Aluminium phosphide
(see hydrogen phos-.
phide)
Amitraz
Amitrole
Azinphos-ethyl
Azinphos-methyl
Azocyclotin
Bendiocarb
Benomyl
BHC (technical grade)
gamma-BHC
Binapacryl
Bioresmethrin
Bitertanol
Bromide ion
Bromomethane
Bromophos
Bromophos-ethyl
Bromopropylate
Butocarboxim
sec-Butylamine
Calcium arsenate
Camphechlor
Captafol
Captan
Carbaryl
Carbendazim
Carbofuran
Carbon disulfide

Carbon tetrachloride
Carbophenothion
Carbosulfan
Cartap
Chinomethionat
Chlorbenside
Chlordane
Chlordimeform
Chlorfenson
Chlorfenvinphos
Chlormequat
Chlorobenzilate
Chloropicrin
Chloropropylate
Chlorothalonil
Chlorpropham
Chlorpyrifos
Chlorpyrifos-methyl
Chlorthion
Coumaphos
Crufomate
Cyanofenphos
Cyhalothrin
Cyhexatin
Cypermethrin
2,4-D
Daminozide
DDT
Deltamethrin
Demeton
Demeton-S-methyl
Demeton-S-methyl
sulfoxide
Demeton-S-methyl
sulfone
Dialifos

Diazinon
1,2-Dibromoethane (see
ethylene dibromide)
1,2-Dichloroethane (see
ethylene dichloride)
Dichlofluanid
Dichloran
Dichlorvos
Dicofol
Dieldrin
Diflubenzuron
Dimethoate
Dimethrin
Dinocap
Disulfoton
Dioxathion
Diphenyl
Diphenylamine
Diquat
Disulfotan
DNOC
Dodine
Edifenphos
Endosulfan
Endrin
Ethephon
Ethiofencarb
Ethion
Ethoprophos
Ethoxyquin
Ethylene dibromide
Ethylene dichloride
Ethylene oxide
Etrimfos
Fenamiphos
Fenbutatin oxide

Fenchlorphos
Fenitrothion
Fensulfothion
Fenthion
Fentin compounds
Fenvalerate
Ferbam
Folpet
Formothion
Guazatine
HCH, technical
Heptachlor + heptachlor
  epoxide
Hexachlorobenzene
  (HCB)
Hydrogen cyanide
Hydrogen phosphide
  (phosphine)
Imazalil
Iprodione
Isofenphos
Lead arsenate
Leptophos
Lindane
Malathion
Maleic hydrazide
Mancozeb
Maneb
Mecarbam
Metalaxyl
Methacrifos
Methamidophos
Methidathion
Methiocarb

Methomyl
Methoprene
Methoxychlor
Methyl bromide
MGK 264
Methyl parathion
Mevinphos
Monocrotophos
Nabam
Nitrofen
Omethoate
Organomercury
  compounds
Oxamyl
Oxydemeton-methyl
Oxythioquinox
Paraquat
Parathion
Parathion-methyl
Permethrin
D-"Phenothrin"
Phenthoate
Phenylmercuric acetate
2-Phenylphenol and its
  sodium salts
Phorate
Phosalone
Phosmet
Phosphamidon
Phosphine (see hydrogen
  phosphide)
Phoxim
Piperonyl butoxide
Pirimucarb

Pirimiphos-methyl
Prochloraz
Procymidone
Propargite
Propamocarb
Propham
Propineb
Propoxur
Pyrethrins
Quintozene
2,4,5-T
Tecnazene
Thiabendazole
Thiofanox
Thiometon
Thiophanate-methyl
Thiram
Toxaphene
Triadimefon
Triazophos
Trichlorfon
Trichloroethylene
Trichloronat
Tricyclohexyltin hydrox-
  ide (see cyhexatin)
Triforine
Triphenyltin
  compounds
Vamidothion
Vinclozolin
Zineb
Ziram

Annex 5

Compounds for which evaluations have been carried out by the Joint FAO/WHO Expert Committee on Food Additives (as of September 1986)

This list includes antioxidants, antibiotics, food colours, contaminants, extraction and carrier solvents, flavouring agents, sweetening agents, xenobiotic anabolic agents and multifunctional compounds. Further information regarding evaluation of specific food additives is available from Division of Environmental Health, World Health Organization, 1211 Geneva 27, Switzerland.

Acesulfame potassium
Acetic acid, glacial
Acetic and fatty acid esters of glycerol
Acetone
Aceton peroxide
Acetylated distarch adipate
Acetylated distarch glycerol
Acetylated distarch phosphate
Acid fuchsine FB
Acrylonitrile
Adipic acid
Agar
Alginic acid
Alkali blue
Alkanet and alkannin
Allura red AC
Allyl-*alpha*-ionone
Allyl-3-cyclohexyl propionate
Allyl hexanoate
Allyl tiglate
Aluminium ammonium sulfate
Aluminium, calcium, magnesium, potassium and sodium salts of caprylic, capric, lauric, and oleic acids
Aluminium potassium sulfate
Aluminium powder
Aluminium silicate
Aluminium sodium sulfate
Aluminium sulfate (anhydrous)

Amaranth
Ammonium acetate
Ammonium adipate
Ammonium alginate
Ammonium carbonate
Ammonium chloride
Ammonium hydrogen carbonate
Ammonium hydroxide
Ammonium lactate
Ammonium persulfate
Ammonium phosphate, dibasic
Ammonium phosphate, monobasic
Ammonium polyphosphate
Ammonium salts of phosphatidic acid
Ammonium stearate
Ammonium succinate
Ampicillin
Amyl acetate
alpha-Amylase (*Aspergillus oryzae*)
alpha-Amyl cinnamic aldehyde
alpha-Amyl cinnamic aldehyde dimethyl acetal
alpha-Amyl cinnamyl alcohol
Amylose, amylopectin
trans-Anethole
Anisyl acetone
Annatto extracts (Bixin and Norbixin)
Anoxomer
Anthocyanin colour from grape skin
Anthocyanins

Arabic gum
Arsenic
beta-Asarone
Asbestos
Ascorbic acid
Ascorbyl palmitate
Ascorbyl stearate
Aspartame
Azodicarbonamide
Azorubine
Bacitracin (*Bacillus subtilis*)
Beet red
Benzaldehyde
Benzene
Benzoic acid
Benzoin gum
Benzoyl superoxide
Benzyl acetate
Benzyl alcohol
Benzyl benzoate
Benzyl butyl ether
Benzyl isobutyl carbinol
Benzyl isoeugenyl ether
Benzyl violet 4B
Bis(2-ethylhexyl) phtalate
Bixin
Black 7984
Blue VRS
Bone phosphate
Borax
Boric acid
Brazil wood
Brilliant Black BN
Brilliant Black PN
Brilliant Blue FCF
Bromelain
Brominated vegetable oils
Brown FK
Brown HT
Butane
1,3-Butane diol
1-Butanol
2-Butanol
Butyl acetate
Butyl butyryl lactate
tert-Butylhydroquinone
Butyl-*p*-hydroxybenzoate
Butylated hydroxyanisole (BHA)
Butylated hydroxytoluene (BHT)
Cadmium
Calcium acetate
Calcium adipate
Calcium alginate

Calcium aluminum silicate
Calcium ascorbate
Calcium benzoate
Calcium carbonate
Calcium chloride
Calcium citrate
Calcium cyclamate
Calcium disodium ethylene-diamine-
  tetraacetate
Calcium ferrocyanide
Calcium fumarate
Calcium gluconate
Calcium di-L-glutamate
Calcium 5'-guanylate
Calcium hydrogen carbonate
Calcium hydrogen phosphate
Calcium hydrogen sulfite
Calcium hydroxide
Calcium iodate
Calcium lactate
Calcium DL-malate
Calcium 5'-inosinate
Calcium metabisulfite
Calcium myristate
Calcium oxide
Calcium palmitate
Calcium peroxide
Calcium phosphate, tribasic
Calcium polyphosphate
Calcium propionate
Calcium pyrophosphate, dibasic
Calcium 5'-ribonucleotides
Calcium saccharin
Calcium silicate
Calcium silicoaluminate
Calcium sorbate
Calcium stearoyl lactylate
Calcium succinate
Calcium sulfate
Canthaxanthin
Capsanthine
Capsorubine
Caramel colour I (plain)
Caramel colour II
Caramel colour III
Caramel colour IV
Carbohydrase from *Aspergillus awa-*
  mori
Carbohydrase from *A. niger*
Carbohydrase from *A. oryzae*
Carbohydrase (alpha amylase) from
  Bacillus lycheniformis
Carbohydrase (pullulanase) from

Klebsiella aerogenes
Carbohydrase from *Rhizopus oryzae*
Carbohydrase from *Saccharomyces* spp.
Carbon, activated vegetable (food grade)
Carbon black
Carbon dioxide
Carboxymethyl cellulose
Carmine: aluminium lake and aluminium calcium lake of carminic acid
Carob bean gum
beta-apo-8'-Carotenal
Carotenes (natural)
beta-Carotene (synthetic)
beta-apo-8'-Carotenic acid, methyl or ethyl ester
Carrageenan (refined non-degraded)
Carthamus red
Carthamus yellow
D-Carvone
L-Carvone
Castor oil
Catalase (bovine liver)
Catalase (*Aspergillus niger*)
Catalase (*Micrococcus lysodeikticus*)
Cellulose, microcrystalline
Cellulose, powdered
Chloramphenicol (*Streptomyces venezuelae*)
Chlorine
Chlorine dioxide
Chloroform
Chlorophyll
Chlorophyll copper complex
Chlorophyllin copper complex, sodium and potassium salts
Chlortetracycline (*Streptomyces aureofaciens*)
Chocolate brown FB
Cholic acid
Choline acetate
Choline carbonate
Choline chloride
Choline citrate
Choline lactate
Choline L(+)-tartrate
Chrysoine
Cinnamaldehyde
Cinnamyl anthranilate
Citral
Citranaxanthin
Citric acid

Citric and fatty acid esters of glycerol
Citronellol (90% and 98% alcohol)
Citrus red No. 2
Cochinol and carminic acid (carmines)
Copper
Coumarin
Crocin and crocetin
Cupric sulfate
Curcumin
Cyclohexane
Cyclohexylsulfamic acid
Dammar gum
Decanal
Desoxycholic acid
Dextrins, white and yellow
Diacetyl
Diacetyltartaric and fatty acid esters of glycerol
Diatomaceous earth
Dibenzyl ether
Dibutyl sebacate
Dichlorodifluoromethane
1,1-Dichloroethane
1,2-Dichloroethane
Dichloromethane (methylene chloride)
Diethyl ether
Diethyl pyrocarbonate
Diethyl tartrate
Diethylene glycol
Diethylene glycol monoethyl ether
Diethylene glycol monopropyl ether
Dihydrostreptomycin (*Streptomyces griseus, S. bikiniensis, S. mashuensis*)
Di-isopropyl ether
Dilauryl thiodipropionate
2,6-Dimethyl-5-heptenal
Dioctyl sodium sulfosuccinate
Diphenyl
Dipotassium hydrogen phosphate
Dipotassium guanylate
Dipropylene glycol
Dipotassium inosinate
Disodium ethylene-diaminetetraacetate
Disodium 5'-guanylate
Disodium hydrogen phosphate
Disodium 5'-inosinate
Disodium pyrophosphate
Disodium 5'-ribonucleotides
Distarch phosphate
Distearyl thiodipropionate
Dodecyl gallate
Dulcin

Eosine

Erythorbic acid (isoascorbic acid)

Erythrosine

Erythromycin (*Streptomyces erythreus*)

Esters of glycerol and thermally oxidized soybean fatty acids

Estragole

Ethanol

Ethyl acetate

Ethyl butyrate

Ethyl cellulose

Ethylenimine

Ethyl formate

Ethyl heptanoate

Ethyl isovalerate

Ethyl *p*-hydroxybenzoate

Ethylhydroxyethyl cellulose

Ethyl lactate

Ethyl laurate

Ethyl maltol

Ethyl methyl ketone

Ethyl methylphenyl glycidate

Ethyl nonanoate

Ethyl 3-phenylglycidate

Ethylene oxide

Ethyl protocatechuate

Ethyl vanillin

Eugenol

Eugenyl methyl ether

Fast green FCF

Fast red E

Fast yellow AB

Ferric ammonium citrate

Ferrous gluconate

Ficin

Food green S

Formic acid

Fumaric acid

Furcelleran

Furfural

Gelatin, edible

Geranyl acetate

Geranyl acetoacetate

Glucono-*delta*-lactone

Glucose isomerase (Immob.) (*Actinoplanes missouriensis*)

Glucose isomerase (*Bacillus coagulans*)

Glucose isomerase (Immob.) (*Bacillus coagulans*)

Glucose isomerase (Immob.) (*Streptomyces olivaceus*)

Glucose isomerase (Immob.) (*S. olivochromogenes*)

Glucose isomerase (*S. rubiginosus*)

Glucose isomerase (Immob.) (*S. rubiginosus*)

Glucose isomerase (*S. violaceoniger*)

Glucose oxidase and catalase (*Aspergillus niger*)

L(+)-Glutamic acid

Glycerol

Glycerol diacetate

Glycerol ester of wood rosin

Glyceryl monoacetate

Gold

Guaiac resin

Guanylic acid

Guar gum

Guinea green B

Gum ghatti

Heptane

Hexamethylenetetraamine

Hexane

alpha-Hexyl cinnamic aldehyde

Hydrochloric acid

Hydrocyanic acid

Hydrogen peroxide

Hydrogenated glucose syrups

Hydroxycitronellal

Hydroxycitronellal dimethyl acetal

Hydroxycitronellol

Hydroxylated lecithin

4-Hydroxymethyl-2,6-di-*tert*-butyl-phenol

Hydroxypropyl cellulose

Hydroxypropyl distarch glycerol

Hydroxypropyl distarch phosphate

Hydroxypropyl methyl cellulose

Hydroxypropyl starch

Indanthrene blue R.S.

Indigotine

Inorganic mercurials

Inosinic acid

Insoluble polyvinyl pyrrolidone (PVPP)

alpha-Ionone

beta-Ionone

Iron

Iron (ferric form)

Iron oxide black

Iron oxide red

Iron oxide yellow

Iron oxides (hydrated and non-hydrated)

Iso-*alpha*-methyl ionone

Isoamyl butyrate

Isoamyl gallate
Isomalt (isomaltitol)
Isopropyl acetate
Isopropyl citrate mixture
Isopropyl myristate
Isoquinoline
Karaya gum
Kanamycin (*Streptomyces kanamyceticus*)
DL-Lactic acid
Lactic and fatty acid esters of glycerol
Lactitol
Lead
Lecithin
Leucomycin (*Streptomyces kitasatoensis*)
Light green SF yellowish
Light petroleum
Linalool, 90% and 95%
Linalyl acetate, 90% and 96%
Lipase, animal
Lipase (*Aspergillus oryzae*)
Lithol rubine BK
Magnesium acetate
Magnesium adipate
Magnesium carbonate
Magnesium chloride
Magnesium citrate
Magnesium gluconate
Magnesium di-L-glutamate
Magnesium hydrogen carbonate
Magnesium hydroxide
Magnesium lactate (DL-)
Magnesium lactate (L-)
Magnesium oxide
Magnesium phosphate, dibasic
Magnesium phosphate, monobasic
Magnesium phosphate, tribasic
Magnesium silicate
Magnesium succinate
DL-Malic acid
Malt carbohydrases
Maltol
Mannitol
Methanol
Menthol (L-and DL-)
Mercury
Methyl anthranilate
Methyl cellulose
alpha-Methyl cinnamic aldehyde
Methyl coumarin
Methyl ethyl cellulose

Methyl *p*-hydroxybenzoate
Methylmercury
Methyl *N*-methylanthranilate
Methyl *beta*-naphthyl ketone
Methyl phenylacetate
2-Methyl-2-propanol
Methyl salicylate
Methyl violet
Mineral oil, food grade
Mixed carbohydrase and protease (*Bacillus subtilis*)
Mixed tartaric, acetic and fatty acid esters of glycerol
Mixed tocopherol concentrate
Monoammonium L-glutamate
Monoammonium orthophosphate
Mono- and diglycerides
Monocalcium phosphate
Monoglyceride citrate
Monomagnesium phosphate, monobasic
Monopotassium L-glutamate
Monosodium L-glutamate
Monosodium monophosphate
Monostarch phosphate
Naphthol yellow S
Natamycin (*Streptomyces natalensis*)
Neomycin (*Streptomyces fradiae*)
Nisin (*Streptococcus lactis*, Lancefield group N)
Nitrogen
2-Nitropropane
Nitrous oxide
gamma-Nonalactone
Nonanal
Norbixin
Nordhydroguaiaretic acid
Novobiocin (*Streptomyces niveus*)
Nystatin (*Streptomyces noursei*)
Oat gum
Octanal
Octyl gallate
Oleandomycin (*Streptomyces antibioticus*)
Orange I
Orange G
Orange GGN
Orange RN
Orchil and orcein
Organomercurial compounds
Organotin compounds
Oxidized hydroxypropyl distarch glycerol

Oxidized starch
Oxystearin
Oxytetracycline (*Streptomyces rimosus*)
Papain
Paprika, oleoresins
Patent blue V
Pectin (amidated)
Pectin (non-amidated)
Penicillins (*Streptomyces* spp.)
Pentapotassium triphosphate
Pentasodium triphosphate
Pepsin
Pepsin, avian
Phenylacetaldehyde
Phenylmercury
o-Phenylphenol
Phosphated distarch phosphate
Phosphoric acid
Pimaricin (*Streptomyces natalensis*)
Piperonal
Polydextrose
Polydimethylsiloxane
Polyethylene glycols
Polyethylenimine
Polyglycerol esters of fatty acids
Polyglycerol esters of interesterified ricinoleic acid
Polymixin B (*Bacillus polymyxa*)
Polyoxyethylene (20) sorbitan mono-laurate
Polyoxyethylene (20) sorbitan mono-oleate
Polyoxyethylene (20) sorbitan mono-palmitate
Polyoxyethylene (20) sorbitan tri-stearate
Polyoxyethylene (8) stearate
Polyoxyethylene (40) stearate
Polyvinylpyrrolidone (PVP)
Ponceau 2R
Ponceau 4R
Ponceau 6R
Ponceau SX
Potassium acetate
Potassium adipate
Potassium alginate
Potassium aluminosilicate
Potassium ascorbate
Potassium benzoate
Potassium bromate
Potassium carbonate
Potassium chlorate
Potassium chloride

Potassium dihydrogen citrate
Potassium dihydrogen phosphate
Potassium ferrocyanide
Potassium fumarate
Potassium gluconate
Potassium hydrogen carbonate
Potassium hydrogen malate
Potassium hydroxide
Potassium iodate
Potassium lactate
DL-Potassium malate solution
Potassium metabisulfite
Potassium nitrate
Potassium nitrite
Potassium persulfate
Potassium phosphate, tribasic
Potassium polyphosphates
Potassium propionate
Potassium saccharin
Potassium sodium tartrate
Potassium sorbate
Potassium succinate
Potassium sulfate
Potassium sulfite
L(+)-Potassium tartrate
Propane
Propan-1-ol
Propan-2-ol
Propionic acid
beta-Propylanisole
p-Propylanisole
Propyl gallate
Propyl p-hydroxybenzoate
1,2-Propylene glycol
1,2-Propylene glycol acetate
1,2-Propylene glycol alginate
Propylene glycol esters of fatty acids
Propylene oxide
Protease (*Aspergillus oryzae*)
Protease (*Streptomyces fradiae*)
Quercetin and quercitron
Quillaia extracts
Quinine hydrochloride
Quinine sulfate
Quinoline yellow (methylated and non-methylated)
Red 10 B
Red 2 G
Rennet
Rennet (bovine)
Rennet (*Bacillus cereus*)
Rennet (*Endothia parasitica*)
Rennet (*Mucor miehei*)

Rennet (*M. pusillus*)
Rennet (*Mucor* spp.)
Rhodiamine B
Riboflavin
Riboflavin 5'-phosphate sodium
Saccharin
Saffron
Safrole and isosafrole
Salicylic acid
Salts of myristic, palmitic, and stearic acid with bases accepted for food use (Al, Ca, Na, Mg, K, ammonium)
Scarlet GN
Silicon dioxide
Silver
Sodium acetate
Sodium adipate
Sodium alginate
Sodium aluminium phosphate, acidic
Sodium aluminium phosphate, basic
Sodium aluminium polyphosphate
Sodium aluminium silicate
Sodium ascorbate
Sodium benzoate
Sodium carbonate
Sodium carboxymethylcellulose
Sodium caseinate
Sodium cyclamate
Sodium diacetate
Sodium dihydrogen citrate
Sodium erythorbate
Sodium ferrocyanide
Sodium fumarate
Sodium gluconate
Sodium hydrogen carbonate
DL-Sodium hydrogen malate
Sodium hydrogen sulfite
Sodium hydroxide
Sodium lactate
DL-Sodium malate
Sodium metabisulfite
Sodium nitrate
Sodium nitrite
Sodium O-phenylphenol
Sodium phosphate, tribasic
Sodium polyphosphate, glassy
Sodium potassium polyphosphate
Sodium propionate
Sodium saccharin
Sodium sesquicarbonate
Sodium sorbate
Sodium stearoyl lactylate

Sodium sulfite
L(+)-Sodium tartrate, monobasic
DL-Sodium tartrate, monobasic
Sodium thiocyanate
Sodium thiosulfate
Sorbic acid
Sorbitan monolaurate
Sorbitan mono-oleate
Sorbitan monopalmitate
Sorbitan monostearate
Sorbitan tristearate
Sorbitol
Sorbitol syrup
Sorboyl palmitate
Spiramycin (*Streptomyces ambofaciens*)
Stannous chloride
Starch acetate
Starch, acid-treated
Starch, alkali-treated
Starch, bleached
Starch, sodium octenyl succinate
Starch, sodium succinate
Starches, enzyme-treated
Stearyoyl monoglyceridyl citrate
Stearyl citrate
Stearyl tartrate
Streptomycin (*Streptomyces griseus*)
Styrene
Succinylated monoglycerides
Sucroglycerides
Sucrose acetate isobutyrate (SAIB)
Sucrose esters of fatty acids
Sudan G
Sudan red G
Sulfur dioxide
Sulfuric acid
Sunset yellow FCF
Talc (magnesium hydrogen metasilicate)
Tannins, food grade
Tara gum
DL-Tartaric acid, ammonium, calcium, and magnesium salts
Tartrazine
Tetrachloroethylene
Tetracycline (*Streptomyces* spp.)
Tetrapotassium pyrophosphate
Tetrasodium diphosphate
Thaumatin
Thiodipropionic acid
Thujone and isothujone
Tin
Titanium dioxide

alpha-Tocopherol
Toluene
Tragacanth gum
Trenbolone acetate
Triacetin
Triammonium citrate
1,1,1-Trichloroethane
1,1,2-Trichloroethylene
1,1,2-Trichlorotrifluoroethane
Triethyl citrate
Triglycerides (synthetic)
Tripotassium citrate
Trisodium citrate
Trypsin

Turmeric
Tylosin (*Streptomyces fradiae*)
Ultramarines
gamma-Undecalactone
Vanillin
Violet 5 BN
Vinyl chloride
Xanthan gum
Xanthophyllis
Xylitol
Yellow 2G
Yellow 27175N
Zeranol
Zinc

Annex 6

Summary of WHO water quality guideline values[a]

Values recommended in these guidelines are for total concentrations (i.e., all forms of substances present).

1. Bacteriological quality

Number of organisms per 100 ml

piped supplies

(*a*) treated water entering the distribution system

faecal coliforms: 0
coliform organisms: 0

(*b*) untreated water entering the distribution system

faecal coliforms: 0
coliform organisms: 0 in 98% of samples over the year; 3 in an occasional sample, but not in consecutive samples.

(*c*) water in the distribution system

faecal coliforms: 0
coliform organisms: 0 in 95% of samples over the year; 3 in an occasional sample, but not in consecutive samples.

unpiped supplies

faecal coliforms: 0
coliform organisms: 10

bottled drinking-water

faecal coliforms: 0
coliform organisms: 0

emergency supplies of drinking-water

faecal coliforms: 0
coliform organisms: 0

---

[a] Further information can be obtained from: WHO. *Guidelines for drinking-water quality*. Geneva, World Health Organization Vol. 1–3. 1984/85.

2. Inorganic constituents of health significance

	Guideline value (mg/litre)
arsenic	0.05
cadmium	0.005
chromium	0.05
cyanide	0.1
fluoride	1.5
lead	0.05
mercury	0.001
nitrate (as N)	10
selenium	0.01

3. Organic constituents of health significance

	Guideline value (µg/litre)
benzene	10
chlorinated alkanes and alkenes	
carbon tetrachloride	3[a]
1,2-dichloroethane	10
1,1-dichloroethene	0.3
tetrachloroethene	10[a]
trichloroethene	30[a]
chlorophenols	
pentachlorophenol	10
2,4,6-trichlorophenol	10
	(odour threshold concentration, 0.1 µg/litre)
polynuclear aromatic hydrocarbons	
benzo[a]pyrene	0.01
trihalomethanes	
chloroform	30
pesticides	
aldrin/dieldrin	0.03
chlordane	0.3
2,4-D	100
DDT	1
heptachlor and heptachlor epoxide	0.1
hexachlorobenzene	0.01
lindane	3
methoxychlor	30

[a] Tentative guideline values. When available carcinogenicity data could not support a guideline value, but the compounds were judged to be of importance in drinking-water and guidance was considered essential, a tentative guideline value was set on the basis of available health-related data.

4. Radioactive materials

gross alpha activity   0.1 Bq/litre
gross beta activity   1 Bq/litre

5. Aesthetic quality

	Guideline value (mg/litre)
aluminium	0.2
chloride	250
copper	1.0
hardness (as $CaCO_3$)	500
iron	0.3
manganese	0.1
sodium	200
sulfate	400
total dissolved solids	1000
zinc	5.0

colour   15 true colour units (TCU)

taste and odour   not offensive to most of the consumers

turbidity   5 nephelometric turbidity units. Preferably < 1 for effective disinfection.

pH   6.5–8.5

Annex 7

Summary of air quality guidelines, developed by the WHO Regional Office for Europe

1. Air quality guideline values for individual substances[a]

Substance	Guideline value	
	Time-weighted average concentration (per m³)	Averaging time
Cadmium	1–5 ng (rural areas)	1 year
	10–20 ng (urban areas)	1 year
Carbon disulfide	100 µg	24 hours
Carbon monoxide	100 mg[b]	15 minutes
	60 mg[b]	30 minutes
	30 mg[b]	1 hour
	10 mg	8 hours
1,2-Dichloroethane	0.7 mg	24 hours
Dichloromethane (methylene chloride)	3 mg	24 hours
Formaldehyde	100 mg	30 minutes
Hydrogen sulfide	150 µg	24 hours
Lead	0.5–1.0 µg	1 year
Manganese	1 µg	1 year[c]
Mercury	1 µg[d] (indoor air)	1 year
Nitrogen dioxide	400 µg	1 hour
	150 µg	24 hours
Ozone	150–200 µg	1 hour
	100–120 µg	8 hours
Styrene	800 µg	24 hours
Sulfur dioxide	500 µg	10 minutes
	350 µg	1 hour
Tetrachloroethene	5 mg	24 hours
Toluene	8 mg	24 hours
Trichloroethene	1 mg	24 hours
Vanadium	1 µg	24 hours

2. Guideline values for combined exposure to sulfur dioxide and particulate matter[a]

Averaging time	Reflectance assessment		Gravimetric assessment	
	SO_2 ($\mu g/m^3$)	Black smoke[b] ($\mu g/m^3$)	Total suspended particles (TSP)[c] ($\mu g/m^3$)	Thoracic particles (TP)[d] ($\mu g/m^3$)
Short term 24 hours	125	125	120[e]	70[e]
Long term 1 year	50	50	—	—

[a] No direct comparisons can be made between values for particulate matter in the right- and left-hand sections of this table, since both the health indicators and the measurement methods differ. While values for total suspended particles and thoracic particles are generally greater than those of black smoke, there is no consistent relationship between them, the ratio of one to the other varying widely depending on the nature of the sources.

[b] Nominal $\mu g/m^3$ units, assessed by reflectance. Application of the black smoke value is recommended only in areas where coal smoke from domestic fires is the dominant component of the particulate matter. It is not necessarily applicable where diesel smoke is an important contributor.

[c] Measurement by high volume sampler, without size selection.

[d] Equivalent values for a sampler with ISO-TP characteristics (having 50% cut-off point at 10 μm): estimated from TSP values using site-specific TSP/ISO-TP ratios.

[e] This value should be regarded as tentative at this stage, since it is based on a single study (involving SO_2 exposure also).

Footnote to table on page 96.

[a] These guideline values were developed by the WHO Regional Office for Europe for use in the European Region. Information from this table should not be used without reference to the rationale given in the full text (WHO. *Air quality guidelines for the European region.* Copenhagen, WHO Regional Office for Europe, 1987).

[b] Exposure at this concentration should be for no longer than the indicated time and should not be repeated within 8 hours.

[c] In view of the respiratory irritancy of this substance, it would be desirable to have a short-term guideline; however, the present data base does not permit such an estimation.

[d] The guideline value is for indoor pollution only: no guideline is given for outdoor concentration that might be of indirect relevance via deposition and entry into the food chain.

3. Guideline values based on sensory effects or annoyance reactions

Substance	Detection threshold (per m^3)	Recognition threshold (per m^3)	Guideline value (per m^3)	Averaging time (minutes)
Carbon disulfide in viscose emissions			20 µg	30
Hydrogen sulfide	0.2–2.0 µg	0.6–6.0 µg	7 µg	30
Styrene	70 µg	210–280 µg	70 µg	30
Tetrachloroethene	8 mg	24–32 mg	8 mg	30
Toluene	1 mg	10 mg	1 mg	30

WHO publications may be obtained, direct or through booksellers, from:

ALGERIA: Entreprise nationale du Livre (ENAL), 3 bd Zirout Youcef, ALGIERS

ARGENTINA: Carlos Hirsch, SRL, Florida 165, Galerías Güemes, Escritorio 453/465, BUENOS AIRES

AUSTRALIA: Hunter Publications, 58A Gipps Street, COLLINGWOOD, VIC 3066 — Australian Government Publishing Service *(Mail order sales)*, P.O. Box 84, CANBERRA A.C.T. 2601 ; *or over the counter from:* Australian Government Publishing Service Bookshops *at:* 70 Alinga Street, CANBERRA CITY A.C.T. 2600 ; 294 Adelaide Street, BRISBANE, Queensland 4000 ; 347 Swanston Street, MELBOURNE, VIC 3000 ; 309 Pitt Street, SYDNEY, N.S.W. 2000 ; Mt Newman House, 200 St. George's Terrace, PERTH, WA 6000 ; Industry House, 12 Pirie Street, ADELAID, SA 5000 ; 156–162 Macquarie Street, HOBART, TAS 7000 — R. Hill & Son Ltd., 608 St. Kilda Road, MELBOURNE, VIC 3004 ; Lawson House, 10–12 Clark Street, CROW'S NEST, NSW 2065

AUSTRIA: Gerold & Co., Graben 31, 1011 VIENNA I

BANGLADESH: The WHO Representative, G.P.O. Box 250, DHAKA 5

BELGIUM: *For books:* Office International de Librairie s.a., avenue Marnix 30, 1050 BRUSSELS. *For periodicals and subscriptions:* Office International des Périodiques, avenue Louise 485, 1050 BRUSSELS — *Subscriptions to World Health only:* Jean de Lannoy, 202 avenue du Roi, 1060 BRUSSELS

BHUTAN: *see* India, WHO Regional Office

BOTSWANA: Botsalo Books (Pty) Ltd., P.O. Box 1532, GABORONE

BRAZIL: Centro Latinoamericano de Informação em Ciencias de Saúde (BIREME), Organização Panamericana de Saúde, Sector de Publicações, C.P. 20381 - Rua Botucatu 862, 04023 SÃO PAULO, SP

BURMA: *see* India, WHO Regional Office

CANADA: Canadian Public Health Association, 1335 Carling Avenue, Suite 210, OTTAWA, Ont. K1Z 8N8. (Tel: (613) 725–3769. Telex: 21–053–3841)

CHINA: China National Publications Import & Export Corporation, P.O. Box 88, BEIJING (PEKING)

DEMOCRATIC PEOPLE'S REPUBLIC OF KOREA: *see* India, WHO Regional Office

DENMARK: Munksgaard Export and Subscription Service, Nørre Søgade 35, 1370 COPENHAGEN K (Tel: + 45 1 12 85 70)

FIJI: The WHO Representative, P.O. Box 113, SUVA

FINLAND: Akateeminen Kirjakauppa, Keskuskatu 2, 00101 HELSINKI 10

FRANCE: Librairie Arnette, 2 rue Casimir-Delavigne, 75006 PARIS

GERMAN DEMOCRATIC REPUBLIC: Buchhaus Leipzig, Postfach 140, 701 LEIPZIG

GERMANY FEDERAL REPUBLIC OF: Govi-Verlag GmbH, Ginnheimerstrasse 20, Postfach 5360, 6236 ESCHBORN — Buchhandlung Alexander Horn, Friedrichstrasse 39, Postfach 3340, 6200 WIESBADEN

GHANA: Fides Enterprises, P.O. Box 1628, ACCRA

GREECE: G.C. Eleftheroudakis S.A., Librairie internationale, rue Nikis 4, ATHENS (T. 126)

HONG KONG: Hong Kong Government Information Services, Beaconsfield House, 6th Floor, Queen's Road, Central, VICTORIA

HUNGARY: Kultura, P.O.B. 149, BUDAPEST 62

INDIA: WHO Regional Office for South-East Asia, World Health House, Indraprastha Estate, Mahatma Gandhi Road, NEW DELHI 110002

INDONESIA: P.T. Kalman Media Pusaka, Pusat Perdagangan Senen, Block 1, 4th Floor, P.O. Box 3433/Jkt, JAKARTA

IRAN (ISLAMIC REPUBLIC OF): Iran University Press, 85 Park Avenue, P.O. Box 54/551, TEHERAN

IRELAND: TDC Publishers, 12 North Frederick Street, DUBLIN 1 (Tel: 744835–749677)

ISRAEL: Heiliger & Co., 3 Nathan Strauss Street, JERUSALEM 94227

ITALY: Edizioni Minerva Medica, Corso Bramante 83–85, 10126 TURIN ; Via Lamarmora 3, 20100 MILAN ; Via Spallanzani 9, 00161 ROME

JAPAN: Maruzen Co. Ltd., P.O. Box 5050, TOKYO International, 100–31

JORDAN: Jordan Book Centre Co. Ltd., University Street, P.O. Box 301 (Al-Jubeiha), AMMAN

KUWAIT: The Kuwait Bookshops Co. Ltd., Thunayan Al-Ghanem Bldg, P.O. Box 2942, KUWAIT

LAO PEOPLE'S DEMOCRATIC REPUBLIC: The WHO Representative, P.O. Box 343, VIENTIANE

LUXEMBOURG: Librairie du Centre, 49 bd Royal, LUXEMBOURG

MALAWI: Malawi Book Service, P.O. Box 30044, Chichiti, BLANTYRE 3

WHO publications may be obtained, direct or through booksellers, from:

MALAYSIA: The WHO Representative, Room 1004, 10th Floor, Wisma Lim Foo Yong (formerly Fitzpatrick's Building), Jalan Raja Chulan, KUALA LUMPUR 05–10; P.O. Box 2550, KUALA LUMPUR 01–02; Parry's Book Center, 124–1 Jalan Tun Sambanthan, P.O. Box 10960, 50730 KUALA LUMPUR

MALDIVES: *See* India, WHO Regional Office

MEXICO: Librería Internacional, S.A. de C.V., Av. Sonora 206, 06100-MÉXICO, D.F.

MONGOLIA: *see* India, WHO Regional Office

MOROCCO: Editions La Porte, 281 avenue Mohammed V, RABAT

NEPAL: *see* India, WHO Regional Office

NETHERLANDS: Medical Books Europe BV, Noorderwal 38, 7241 BL LOCHEM

NEW ZEALAND: New Zealand Government Printing Office, Publishing Administration, Private Bag, WELLINGTON; Walter Street, WELLINGTON; World Trade Building, Cubacade, Cuba Street, WELLINGTON, *Government Bookshops at:* Hannaford Burton Building, Rutland Street, Private Bag, AUCKLAND; 159 Hereford Street, Private Bag, CHRISTCHURCH; Alexandra Street, P.O. Box 857, HAMILTON; T & G Building, Princes Street, P.O. Box 1104, DUNEDIN — R. Hill & Son Ltd, Ideal House, Cnr Gillies Avenue & Eden Street, Newmarket, AUCKLAND 1

NORWAY: Tanum — Karl Johan A.S., P.O. Box 1177, Sentrum, N-0107 OSLO 1

PAKISTAN: Mirza Book Agency, 65 Shahrah–E–Quaid–E–Azam, P.O. Box 729, LAHORE 3

PAPUA NEW GUINEA: The WHO Representative, P.O. Box 646, KONEDOBU

PHILIPPINES: World Health Organization, Regional Office for the Western Pacific, P.O. Box 2932, MANILA

PORTUGAL: Livraria Rodrigues, 186 Rua do Ouro, LISBON 2

REPUBLIC OF KOREA: The WHO Representative, Central P.O. Box 540, SEOUL

SINGAPORE: The WHO Representative, 144 Moulmein Road, SINGAPORE 1130; Newton P.O. Box 31, SINGAPORE 9122

SOUTH AFRICA: *Contact major book stores*

SPAIN: Ministerio de Sanidad y Consumo, Centro de publicaciones, Documentación y Biblioteca, Paseo del Prado 18, 28014 MADRID — Comercial Atheneum S.A., Consejo de Ciento 130–136, 08015 BARCELONA; General Moscardó 29, MADRID 20 — Librería Díaz de Santos, P.O. Box 6050, 28006 MADRID; Balmes 417 y 419, 08022 BARCELONA

SRI LANKA: *see* India, WHO Regional Office

SWEDEN: *For books:* Aktiebolaget C.E. Fritzes Kungl. Hovbokhandel, Regeringsgatan 12, 103 27 STOCKHOLM. *For periodicals:* Wennergren-Williams AB, Box 30004, 104 25 STOCKHOLM

SWITZERLAND: Medizinischer Verlag Hans Huber, Länggassstrasse 76, 3012 BERN 9

THAILAND: *see* India, WHO Regional Office

UNITED KINGDOM: H.M. Stationery Office: 49 High Holborn, LONDON WC1V 6HB; 13a Castle Street, EDINBURGH EH2 3AR, 80 Chichester Street, BELFAST BT1 4JY; Brazennose Street, MANCHESTER M60 8AS; 258 Broad Street, BIRMINGHAM B1 2HE; Southey House, Wine Street, BRISTOL BS1 2BQ. *All mail orders should be sent to:* HMSO Publications Centre, 51 Nine Elms Lane, LONDON SW8 5DR

UNITED STATES OF AMERICA: *Copies of individual publications (not subscriptions):* WHO Publications Center USA, 49 Sheridan Avenue, ALBANY, NY 12210. *Subscription orders and correspondence concerning subscriptions should be addressed to the* World Health Organization, Distribution and Sales, 1211 GENEVA 27, Switzerland. *Publications are also available from the* United Nations Bookshop, NEW YORK, NY 10017 (*retail only*)

USSR: *For readers in the USSR requiring Russian editions:* Komsomolskij prospekt 18, Medicinskaja Kniga, MOSCOW — *For readers outside the USSR requiring Russian editions:* Kuzneckij most 18, Meždunarodnaja Kniga, MOSCOW G-200

VENEZUELA: Librería Medica Paris, Apartado 60.681, CARACAS 106

YUGOSLAVIA: Jugoslovenska Knjiga, Terazije 27/II, 11000 BELGRADE

Special terms for developing countries are obtainable on application to the WHO Representatives or WHO Regional Offices listed above or to the World Health Organization, Distribution and Sales Service, 1211 Geneva 27, Switzerland. Orders from countries where sales agents have not yet been appointed may also be sent to the Geneva address, but must be paid for in pounds sterling, US dollars, or Swiss francs. Unesco book coupons may also be used.

Prices are subject to change without notice.